*Applied Probability and Statistics*

BAILEY · The Elements of Stochastic Processes with Applications to the Natural Sciences

BAILEY · Mathematics, Statistics and Systems for Health

BARTHOLOMEW · Stochastic Models for Social Processes, *Second Edition*

BECK and ARNOLD · Parameter Estimation in Engineering and Science

BENNETT and FRANKLIN · Statistical Analysis in Chemistry and the Chemical Industry

BHAT · Elements of Applied Stochastic Processes

BLOOMFIELD · Fourier Analysis of Time Series: An Introduction

BOX · The Life of a Scientist

BOX and DRAPER · Evolutionary Operation: A Statistical Method for Process Improvement

BOX, HUNTER, and HUNTER · Statistics for Experimenters: An Introduction to Design, Data Analysis and Model Building

BROWN and HOLLANDER · Statistics: A Biomedical Introduction

BROWNLEE · Statistical Theory and Methodology in Science and Engineering, *Second Edition*

BURY · Statistical Models in Applied Science

CHAMBERS · Computational Methods for Data Analysis

CHATTERJEE and PRICE · Regression Analysis by Example

CHERNOFF and MOSES · Elementary Decision Theory

CHOW · Analysis and Control of Dynamic Economic Systems

CLELLAND, deCANI, BROWN, BURSK, and MURRAY · Basic Statistics with Business Applications, *Second Edition*

COCHRAN · Sampling Techniques, *Third Edition*

COCHRAN and COX · Experimental Designs, *Second Edition*

COX · Planning of Experiments

COX and MILLER · The Theory of Stochastic Processes, *Second Edition*

DANIEL · Application of Statistics to Industrial Experimentation

DANIEL and WOOD · Fitting Equations to Data

DAVID · Order Statistics

DEMING · Sample Design in Business Research

DODGE and ROMIG · Sampling Inspection Tables. *Second Edition*

DRAPER and SMITH · Applied Regression Analysis

DUNN · Basic Statistics: A Primer for the Biomedical Sciences, Second Edition

DUNN and CLARK · Applied Statistics: Analysis of Variance and Regression

ELANDT-JOHNSON · Probability Models and Statistical Methods in Genetics

FLEISS · Statistical Methods for Rates and Propor

GIBBONS, OLKIN and SOB ula-tions: A New Statistical M

GNANADESIKAN · Metho Multivariate Observations

GOLDBERGER · Economet

GROSS and CLARK · Survi ca-tions in the Biomedical Scie

GROSS and HARRIS · Fundamentals of Queueing Theory

GUTTMAN, WILKS and HUNTER · Introductory Engineering Statistics, *Second Edition*



*continued on back*

# Foundations of Inference in Survey Sampling

# Foundations of Inference in Survey Sampling

**CLAES-MAGNUS CASSEL**

The National Central Bureau of Statistics,
Stockholm

**CARL-ERIK SÄRNDAL**

The University of British Columbia,
Vancouver

**JAN HÅKAN WRETMAN**

The National Central Bureau of Statistics,
Stockholm

A WILEY–INTERSCIENCE PUBLICATION

JOHN WILEY & SONS, NEW YORK · LONDON · SYDNEY · TORONTO

**Library of Congress Cataloging in Publication Data:**

Cassel, Claes-Magnus, 1940–
Foundations of inference in survey sampling.

(Wiley series in probability and mathematical statistics)
Bibliography: p.
Includes index.
1. Sampling (Statistics) 2. Estimation theory
I. Särndal, Carl-Erik, 1937– joint author
II. Wretman, Jan Håkan, 1939– joint author.
III. Title
QA276.6.C34 519.5′2 77-5114
ISBN 0-471-02563-1

Printed in the United States of America

10 9 8 7 6 5 4 3 2 1

# Preface

Until recently survey sampling and statistical inference theory were surprisingly distinct fields, though they ought to have a lot in common. The practical problems of survey sampling did not challenge the inference theorists, and inference theory often seemed too abstract to help the survey samplers.

The last two decades have brought a change for which credit is due to statisticians who combined strong interests in the two fields. Consequently, the basic theory of survey sampling is in a state of reevaluation and lively development.

An early well-known result in this trend (Godambe, 1955) states that no uniformly minimum variance estimator of the finite population mean exists, within a certain (reasonable) class of estimators. This and other unconventional results served as eyeopeners. Even basic and widely used estimators, such as the sample mean, were suddenly cast in doubt. Precisely what favorable properties, if any, could be claimed for the estimators usually relied on in sampling practice?

Many traditional sampling techniques had proven their value from a practical point of view, but had remained ad hoc procedures from a statistical inference point of view. The new trend in survey sampling has contributed a more satisfactory apparatus for evaluating the traditional techniques. The conditions under which a given procedure is optimal have been explored in detail.

The debate on the problems of survey sampling has raised issues of broad interest in statistical inference. Various concepts in statistical inference theory have been applied in survey sampling with mixed results. For example, why is the likelihood method essentially a failure in the survey sampling situation? How should one explain the ambiguous role of "man-made randomization," traditionally imposed in survey sampling in the form of a sampling design?

Moreover, certain practices typical to survey sampling have raised important questions which remain to be answered by statistical theory. For example, if observations have been sampled according to "probability proportional to size," rather than at random, how valid are traditional methods in the theory of hypothesis testing, or in any of the established modes of statistical inference?

The Superpopulation approach is one avenue through which important new methods are currently being added to the survey sampler's traditional set of tools. This promising approach, which occupies a major portion of our text, contributes significantly towards a better understanding of a variety of survey sampling problems.

Our purpose in writing this text has been to provide the first reasonably complete account of what might be called the statistical inference theory outlook on survey sampling.

Considerable effort was expended in developing a flexible system of notation. As far as the material is concerned, we have selected and emphasized certain topics that seemed of particular importance among the many contributions to finite population inference over the past 15–20 years.

Naturally, the selection reflects our own biases as to what is important to the basic theory. Some results have the character of hitherto unpublished research.

This is not a text on the pragmatics of how to run a survey, nor is it like a traditional text on sampling theory.

If we look at a "classical" one of the latter kind, such as W. G. Cochran: *Sampling Techniques*, we shall hardly see mentioned those concepts of inference, for example, sufficiency, likelihood, admissibility, Rao–Blackwellization, which are essential for an understanding of the present text.

Our text will not replace sampling texts such as Cochran's and others of similar type. We attempt to examine the subject from quite a different angle.

We have aimed at a level where advanced knowledge of theory should not be a necessary prerequisite: the theory of statistics content is rather of an "intermediate" character. Familiarity with traditional concepts and approaches in statistical inference is assumed, for example, at the level of B. W. Lindgren: *Statistical Theory*.

In addition, a background in survey sampling theory and method, for example, according to the already mentioned book by Cochran, is certainly helpful for a full appreciation of the material.

We hope that our text will find application both as a self-contained course in the foundations of inference for finite populations and as an introduction to research in the area.

The book has two parts: the Fixed population approach (Chapters 2–3) and the Superpopulation approach (Chapters 4–7).

Each chapter is divided into sections. Formulas are numbered section by section, without chapter indication. For example, in any given chapter, formula (4.2) refers to formula no. 2 in Section 4 of the chapter. Numbering of lemmas, theorems, and corollaries, and tables is, on the other hand, on a chapter basis. For example, Theorem 3.2 refers to the second theorem of Chapter 3. In addition, each chapter contains a number of definitions, remarks, and examples. A remark serves to give emphasis to important subject matter, for example, relating to a theorem just proven.

The project of authoring this book has received financial backing from the Swedish Council for Social Science Research and from the National Research Council of Canada. We also gratefully acknowledge the cooperation of our respective employers, The National Central Bureau of Statistics, Stockholm, and The University of British Columbia, Vancouver, especially in providing us with time off from our regular duties during various phases of the project.

This book is the result of complete teamwork, and it is impossible to single out a "senior author." Therefore we have chosen to let our names appear in alphabetical order on the cover and title page.

C. M. CASSEL
C. E. SÄRNDAL
J. H. WRETMAN

*Stockholm and Vancouver*
*February 1977*

# Contents

# Foundations of Inference in Survey Sampling

CHAPTER 1

# Basic Model of Sampling from a Population with Identifiable Units

The essence of survey sampling consists of the selection of a part of a finite collection of units, followed by the making of statements about the entire collection on the basis of the selected part. In this process the statistician must be guided by a body of principles and procedures for statistical inference. His analysis may involve several sources of randomness, for example,

(i) the method by which units were selected,
(ii) the method used to obtain a measurement for a selected unit,
(iii) knowledge of some process that generates the true measurement for a given unit.

Each of these elements would contribute stochastic structure to the inference.

## 1.1. FIXED POPULATION APPROACH VERSUS SUPERPOPULATION APPROACH

To be more specific, case (i) above refers to the stochastic element introduced whenever sampling is nonpurposive, that is, when a randomizing mechanism determines what part of the collection to observe.

Once a unit has been selected for the sample, it remains to obtain a measurement on the target variable. Case (ii) refers to the situation where such measurement is imperfect, that is, there exist errors for which one may nevertheless be able to pose a stochastic model.

Thirdly, if a given unit is part of a survey on two separate occasions, say, one year apart, then a minimal expenditure of effort may produce a highly accurate forecast for this unit the second time around. It may suffice to know last year's known value, plus a stochastic model for how that value relates to this year's unknown value; this is an example of the kinds of models referred to in case (iii) above.

We shall consider statistical inference procedures for finite collections of units where stochastic elements are introduced through the possibilities (i): the randomization aspect, and (iii): the stochastic model, or Superpopulation, aspect. A detailed treatment of the possibility (ii): models for measurement errors, or, more generally, the theory for nonsampling errors, will be considered beyond the scope of this text; a reference in this area is Hansen, Hurwitz and Pritzker (1963).

As an example, suppose that the units are farms and that the characteristic under study is the yield of wheat in a given year. One approach to the inference problem, often used in standard texts, is the following: It is assumed that to each farm in the population corresponds a fixed but unknown real number representing the yield of that particular farm in the particular year. When a farm has been selected for the sample, it is furthermore assumed that the fixed real number corresponding to this farm is measured without error. These assumptions will be made in the early chapters of this book.

A different approach to the problem is to treat the yields of the farms in the population as numbers generated under a stochastic model. Such models often incorporate auxiliary knowledge.

A crude but frequently effective model simply postulates a linear stochastic relationship, for example, that yield of wheat is, apart from an error term of zero expected value, proportional to size of the farm in acres, $x_k$, which quantity is assumed known from a previous year. Moreover, the model considers that the unknown proportionality factor $\beta$ is common to all farms. It would be determined, in the particular year, by the average propensity of farmers to devote acreage to wheat, by average yield per acre that year, and so on.

An estimate $\hat{\beta}$ of the unknown proportionality factor can be obtained from a sample of farms. For any one farm *not* in the sample, the value $\hat{\beta}x_k$ should provide an effective prediction of wheat yield, thereby permitting a prediction of total yield in the population.

By the above discussion we have indicated the basic dichotomy of the material to be presented in this book:

(i) *The Fixed population approach:* With each population unit is associated a fixed but unknown real number, that is, the value of the variable under study.

(ii) *The Superpopulation approach:* With each population unit is associated a random variable for which a stochastic structure is specified; the actual value associated with a population unit is treated as the outcome of this random variable.

The choice of approach naturally depends on the situation at hand. The Fixed population approach is the "traditional" one in survey sampling. However, there are also early examples of the Superpopulation approach: Cochran (1939, 1946), Deming and Stephan (1941). Many recent important contributions to finite population inference theory take the Superpopulation approach, which is an avenue of great promise in survey sampling.

In the first part of this book (Chapters 2–3), we discuss contributions to the theory of finite population inference under the Fixed population approach. In the second part (Chapters 4–7), the topic is treated through the Superpopulation approach. The two approaches build on a common basic framework to be introduced in this chapter. Suitable extensions of this framework will be given in Chapter 4 in order to handle the Superpopulation approach.

## 1.2. THE BASIC MODEL

Occasionally we shall make comparisons between survey sampling inference and traditional statistical inference, by which we shall mean inference for the infinite, "hypothetical" populations usually assumed in statistical theory since the times of R. A. Fisher. In that setup there is typically a sample of $n$ independent observations $x_1, \ldots, x_n$ on a random variable $X$ with the hypothetical density function $f(x;\theta)$, and the problem is often to estimate the unknown parameter $\theta$. Some survey sampling theorists, for example, Godambe (1970), like to emphasize the hypothetical nature of such populations, as compared to the "real" populations in survey sampling.

Traditional statistical inference and survey sampling inference are not opposing theories, but the special nature of the latter produces some unexpected results. The basic concepts, such as parameter, sample, data, estimator, are given a special meaning in survey sampling. We suggest that the reader used to the framework of traditional statistical inference make particular note of the definitions (given in this chapter) of these basic concepts. The notion of sampling design should also be noted; this concept has no direct counterpart in traditional statistical theory, although in the area of experimental design we find the related idea of manipulation of experimental conditions as a means of gaining better inferences.

We shall now define the basic model of survey sampling underlying this book. The basic model incorporates certain fundamental features of sampling from finite populations. An essential ingredient is the identifiability of units. This feature, usually missing in traditional statistical theory, is in fact responsible for some of the unexpected results we find in survey sampling inference.

The identifiability means, for example, that the statistician is free to choose his own sampling design; that is, "man-made randomization" is used in selecting a sample. The sampling distribution of a given estimator is thereby something that the statistician creates. Thus in survey sampling we are not confined to independent and identically distributed observations, as is often the case in traditional statistical inference. Also, in survey sampling, the sample mean is not of dominating importance as an estimator of the population mean. We have in fact a variety of interesting possibilities for estimating this population parameter.

The formulation of the basic model is due essentially to Godambe (1955), with later refinements by Godambe and others. The basic model has become the widely accepted point of departure by writers on the topic of finite population inference; see, for example, the overview papers by Basu (1971), J. N. K. Rao (1975), Solomon and Zacks (1970).

We shall now describe the model as it will be used in the Fixed population approach. The model will then be suitably extended in Chapter 4 to accommodate the Superpopulation approach.

**Definition.** *A finite population* is a collection of $N$ units, where $N < \infty$, and $N$ is called the *size of the population*. ■

An important assumption made throughout the book is that $N$ is known.

This book deals exclusively with finite populations whose units are identifiable. Without going into details about the nature of the property (or the properties) that makes it possible to tell the units apart, let us just assume that identification is possible, and that identity is indicated by means of an integer, $k$, called *the label* of unit $u_k$. For example, $k$ could simply be the number giving the order of appearance in a complete listing of the population.

**Definition.** The units of a finite population are said to be *identifiable* if they can be uniquely labeled from 1 to $N$ and the label of each unit is known. ■

We shall often use the label $k$ to represent the physically existing unit $u_k$. We shall talk about $u_k$ as "the unit $k$" and thus denote the population as

$$\mathcal{U} = \{1, \ldots, k, \ldots, N\}$$

With unit $k$ is associated a number $y_k$ when the characteristic of interest is $y$. In the abstract sense, the symbol $y_k$ denotes "the value that $y$ takes for the given unit $k$." We assume away measurement error: As soon as unit $k$ is included in our sample, we can measure $y_k$ without error. As already pointed out, the statistical theory dealing with the nonsampling error arising when $y_k$ cannot be exactly observed is beyond the scope of this text. (Our framework could, of course, be extended to incorporate nonsampling error.)

In the practical sense, $y_k$ is, however, a real number, the result of measuring unit $k$. If measurement of $y$ for unit 2 gave the value 10, then $y_2 = 10$, but from the number 10 alone we have no way of telling which unit was under observation. Since labels are important, a more precise notation is needed.

When unit $k$ has been measured, we should record not only the number $y_k$, but also the fact that it was precisely unit $k$ that produced that measurement. We denote "the complete observation" or "the labeled observation" by the pair $(k, y_k)$. (This detailed notation is out of question in traditional statistical inference, since labels are lacking.)

**Example 1.** Assume that we have recorded the set of pairs $\{(1,28), (2,20)\}$, that is, unit 1 with the value $y_1 = 28$, and unit 2 with the value $y_2 = 20$.

If we were to suppress the labels, we are left with the set of numbers $\{20, 28\}$. But because the information about the labels is there, we cannot arbitrarily decide to ignore it before making inference.

Assume that for unit 3 we have $(3, y_3) = (3, 20)$. The information $\{(1, 28), (3, 20)\}$ based on units 1 and 3 does not necessarily give the same inference as $\{(1, 28), (2, 20)\}$ based on units 1 and 2, even though the set of unlabeled measurements, $\{20, 28\}$, is the same.

As an example of this, assume that the units 1, 2 and 3 define the entire population, that is, $N = 3$. Let $p(s_1) = 0.32$, $p(s_2) = 0.40$, $p(s_3) = 0.28$ be the probabilities with which the label sets $s_1 = \{1, 2\}$, $s_2 = \{1, 3\}$, and $s_3 = \{2, 3\}$ are selected, respectively. The estimator $t$ defined by

$$t = \begin{cases} (y_1 + y_2)/6\,p(s_1) & \text{if} \quad s_1 \quad \text{occurs} \\ (y_1 + y_3)/6\,p(s_2) & \text{if} \quad s_2 \quad \text{occurs} \\ (y_2 + y_3)/6\,p(s_3) & \text{if} \quad s_3 \quad \text{occurs} \end{cases}$$

is unbiased (see Section 1.6) for the population mean $(y_1 + y_2 + y_3)/3 = 22.67$.

In this example it would be hard to find a reason as to why one should choose these particular probabilities for the label sets $s_i (i = 1, 2, 3)$. However we shall see later that there are strong reasons for working with unequal probability sampling.

If $\{1, 2\}$ obtains, the labeled data are $\{(1, 28), (2, 20)\}$, and the estimate is $t = (28 + 20)/6 \times 0.32 = 25.0$. By contrast if $\{1, 3\}$ obtains, the labeled data are $\{(1, 28), (3, 20)\}$, so the estimate is now $t = (28 + 20)/6 \times 0.40 = 20.0$.

If on the other hand we had chosen a sample by the probabilities $p(s_1) = p(s_2) = p(s_3) = 1/3$, as in simple random sampling, then the estimate is indeed the same for both $s_1$ and $s_2$, $t = (28 + 20)/6 \times 1/3 = 24.0$. That is, the distribution of the estimator has changed.

The formal distinction between "labeled data" and "unlabeled data" will be introduced in Section 1.5. ■

Now, for the whole population we have a set of $N$ labeled pairs, $(k, y_k)$ $(k = 1, \ldots, N)$. It is natural to identify a parameter for the finite population in the following way:

**Definition.** The vector $\mathbf{y} = (y_1, \ldots, y_N)$ is called a *parameter* of the finite population. ■

**Remark 1.** We shall use the convention that the $k$th component of the vector is associated with unit $k$. Thus if $\mathbf{y}$ is known, the components of $\mathbf{y}$ are in effect labeled by their position in the vector. ■

In traditional statistical inference a parameter is frequently treated as an unknown constant indexing a probability distribution. In Section 2.1 we shall see that $\mathbf{y}$ is indeed a parameter in this sense in the case of finite populations, too.

We shall denote the parameter space by $\Omega$. Frequently, $\Omega = R_N$, the $N$-dimensional Euclidean space. Occasionally, other parameter spaces of practical importance will be considered, for example, the case where each component $y_k$ of $\mathbf{y}$ is limited to taking the values 0 or 1.

Any real function of $y_1, \ldots, y_N$ is called a *parametric function*. Inference in finite populations is usually about a specific parametric function and seldom about $\mathbf{y}$ itself.

Two important parametric functions are the *population mean*,

$$\bar{y} = \frac{\Sigma_1^N y_k}{N}$$

and the *population variance*,

$$\sigma_y^2 = \frac{\Sigma_1^N (y_k - \bar{y})^2}{N}$$

**Remark 2.** In summing over the population units, we shall write $\Sigma_1^N$, $\Sigma\Sigma_1^N$, and $\Sigma\Sigma_{k \neq l}^N$, respectively, for

$$\sum_{k=1}^{N}, \sum_{k=1}^{N}\sum_{l=1}^{N} \quad \text{and} \quad \sum_{\substack{k=1 \\ k \neq l}}^{N}\sum_{l=1}^{N}$$ ■

We shall study principles for making statistical inference about $\bar{y}$, based on information obtained from a part of $\mathscr{U}$. We introduce first the concept of a sample.

**Definition.** A sequence $\mathbf{s} = (k_1, \ldots, k_{n(\mathbf{s})})$ such that $k_i \in \mathscr{U}$ for $i = 1, \ldots, n(\mathbf{s})$ is called an *ordered sample*, or simply a sample. The label $k_i$ is called the $i$th *component* of $\mathbf{s}$. The number of components of $\mathbf{s}$, $n(\mathbf{s})$, is called the *sample size*. ■

The set of all sequences $\mathbf{s}$ will be denoted by $\mathscr{S}^*$.

In an ordered sample s, two or more components may be identical. A unit occurring more than once in s will be called a *repeat*. In principle, $n(\mathbf{s})$ may exceed $N$. If repeats are allowed, $\mathcal{S}^*$ is at most a countable set.

**Definition.** The number of distinct components of a sequence s will be denoted $\nu(\mathbf{s})$ and called the *effective sample size*. ■

**Remark 3.** Somewhat improperly, we shall use a simplifying piece of notation usually reserved for sets, namely, $k \in \mathbf{s}$, to indicate that $k$ is a component of s. Also, $k \in \mathbf{s}$ will indicate enumeration of the components of s. When summing over the units $k = k_i$, $i = 1, \ldots, n(\mathbf{s})$, we shall write $\Sigma_\mathbf{s}$. For example, $\Sigma_\mathbf{s} y_k$ will stand for

$$\sum_{i=1}^{n(\mathbf{s})} y_{k_i}$$ ■

The definition of a sequence as a sample is obviously tied to the idea that $n(\mathbf{s})$ consecutive "draws" are made from the population, with or without replacement, such that $i$ represents drawing order, and such that $k_i$ is the label resulting in the $i$th draw.

**Example 2.** Let $\mathcal{U} = \{1, 2, 3\}$. If repeats are not allowed, $n(\mathbf{s}) = \nu(\mathbf{s})$ for all s and we have the following 15 samples:

$$\begin{array}{lll}
\mathbf{s}_1 = (1) & \mathbf{s}_6 = (1, 3) & \mathbf{s}_{11} = (2, 3, 1) \\
\mathbf{s}_2 = (2) & \mathbf{s}_7 = (3, 1) & \mathbf{s}_{12} = (3, 1, 2) \\
\mathbf{s}_3 = (3) & \mathbf{s}_8 = (2, 3) & \mathbf{s}_{13} = (1, 3, 2) \\
\mathbf{s}_4 = (1, 2) & \mathbf{s}_9 = (3, 2) & \mathbf{s}_{14} = (2, 1, 3) \\
\mathbf{s}_5 = (2, 1) & \mathbf{s}_{10} = (1, 2, 3) & \mathbf{s}_{15} = (3, 2, 1)
\end{array}$$ ■

If we are interested only in which units are included in the sample s, leaving aside their drawing order as well as information about repeats, we can define a set $s$ corresponding to the sequence s:

$$s = \{k : k \text{ is a component of } \mathbf{s}\}$$

Any repeats in s are represented once and only once in $s$. The number of elements in $s$ is therefore $\nu(\mathbf{s})$.

If $s$ is any set, we shall also write $\nu(s)$ to denote the number of elements in $s$, which by definition are distinct. In particular if $s$ is the set derived by "reducing" a sequence s, then $\nu(s) = \nu(\mathbf{s})$. No ambiguity should derive from this notation, although we are dealing with two different functions $\nu(\cdot)$, one being a function of a set and the other of a sequence.

**Definition.** A nonempty set $s$ such that $s \subseteq \mathcal{U}$ is called an *unordered sample*. The number of elements of $s$, $\nu(s)$, is called the *effective sample size*. ■

("Sample size" and "effective sample size" would be identical quantities for a set.)

The set of all sets $s$ will be denoted by $\mathscr{S}$.

**Remark 4.** When summing over $k$ in $s$ we shall for simplicity write $\Sigma_s$, for example, $\Sigma_s y_k$ will stand for $\Sigma_{k \in s} y_k$. ■

To make more explicit the fact that a set $s$ may obtain by "reducing" a sequence $\mathbf{s}$, we introduce the *reducing function* $r(\mathbf{s})$, a mapping of $\mathscr{S}^*$ onto $\mathscr{S}$: For every $\mathbf{s} \in \mathscr{S}^*$ there is an $s \in \mathscr{S}$ such that $r(\mathbf{s}) = s$.

**Example 3.** Let $\mathscr{U} = \{1, 2, 3\}$. In Example 2, the set of sequences was $\{\mathbf{s}_1, \ldots, \mathbf{s}_{15}\}$. Moreover, $\mathscr{S} = \{s_1, \ldots, s_7\}$, where $s_1 = \{1\}$, $s_2 = \{2\}$, $s_3 = \{3\}$, $s_4 = \{1, 2\}$, $s_5 = \{1, 3\}$, $s_6 = \{2, 3\}$, $s_7 = \{1, 2, 3\}$. Thus, the function $r(\mathbf{s})$ is such that $r(\mathbf{s}_i) = s_i$ for $i = 1, 2, 3$; $r(\mathbf{s}_i) = s_4$ for $i = 4, 5$; $r(\mathbf{s}_i) = s_5$ for $i = 6, 7$; $r(\mathbf{s}_i) = s_6$ for $i = 8, 9$; and $r(\mathbf{s}_i) = s_7$ for $i = 10, 11, \ldots, 15$. ■

Our first definition of a sample was in terms of a sequence $\mathbf{s}$. However, for reasons to be explained in Chapter 2, we can often without loss of essential information limit interest to the resulting set, $s = r(\mathbf{s})$.

This defines one out of two ways in which we can consider that a set $s$ is brought to the attention of the statistician responsible for making the inference:

(i) The statistician starts out with a sequence $\mathbf{s}$ and reduces it to the corresponding set $s = r(\mathbf{s})$. This way he knows both $\mathbf{s}$ and $s = r(\mathbf{s})$.

The second way is the following:

(ii) The statistician knows only the set $s$. One possibility is that someone else has reduced the sequence $\mathbf{s}$ into the set $s$, without providing the statistician with the information $\mathbf{s}$. Another possibility is that $s$ was drawn as a set in the first place, that is, by "mass draw" of the $\nu(s)$ units from $\mathscr{U}$.

In the following we shall make a distinction between discussion relating to sequences $\mathbf{s}$ on the one hand and relating to sets $s$ on the other. We shall talk about "the sequence case" and "the set case," respectively.

**Remark 5.** When summation is over a specified set $A$ of sequences $\mathbf{s}$, or of sets $s$, we shall write simply $\Sigma_A$. For example, if $p(\cdot)$ is a given function on $\mathscr{S}^*$, or on $\mathscr{S}$, we shall write $\Sigma_A\, p(\mathbf{s})$ in the sequence case and $\Sigma_A\, p(s)$ in the set case. ■

## 1.3. SAMPLING DESIGNS: GENERAL PROPERTIES

We now introduce the important concept of a sampling design as a function $p(\cdot)$ on $\mathscr{S}^*$ or on $\mathscr{S}$.

Through a sampling design, "unequal probability sampling" becomes a possibility in surveys. However, the role of sampling design in survey sampling inference is not clear. In classical survey sampling it is an all-important element, whereas adherents of newer, superpopulation based inference claim that the way in which the actual sample was selected is unimportant. Chapters 2–4 deal with design-oriented inference, while Chapters 5–7 are superpopulation oriented.

A distinction will be made between ordered and unordered designs, as follows:

To each $\mathbf{s} \in \mathscr{S}^*$, attach a real number $p(\mathbf{s})$ such that

$$p(\mathbf{s}) \geqslant 0 \quad \text{for all} \quad \mathbf{s} \in \mathscr{S}^* \tag{3.1}$$

and

$$\Sigma_{\mathscr{S}^*} p(\mathbf{s}) = 1 \tag{3.2}$$

**Definition.** A function $p(\mathbf{s})$ on $\mathscr{S}^*$ satisfying (3.1) and (3.2) will be called an *ordered sampling design.* ∎

Let $\mathbf{S}$ be the random variable taking values $\mathbf{s} \in \mathscr{S}^*$. The design $p(\mathbf{s})$ then describes the discrete probability distribution of $\mathbf{S}$:

$$P(\mathbf{S} = \mathbf{s}) = p(\mathbf{s}) \quad \text{for} \quad \mathbf{s} \in \mathscr{S}^*$$

Let $K_i$ be the random variable associated with the $i$th position of $\mathbf{S}$ (the $i$th draw), such that $K_i$ takes the value $k_i$ if unit $k_i$ results in the $i$th draw. Hence, the random variable $\mathbf{S}$ can be written as

$$\mathbf{S} = (K_1, \ldots, K_{n(\mathbf{S})})$$

In the set case, let $S$ denote the random variable taking values $s \in \mathscr{S}$. In case (i) of Section 1.2, the statistician would usually know the distribution, $p(\mathbf{s})$ of $\mathbf{S}$. He can then obtain the induced distribution of $S$ as

$$P(S = s) = \Sigma_{A_s} p(\mathbf{s})$$

where

$$A_s = \{\mathbf{s} : r(\mathbf{s}) = s\}$$

In case (ii) of Section 1.2, the statistician would usually know the distribution $p(s)$ of $S$.

We shall write $P(S = s) = p(s)$, although strictly speaking $p(\mathbf{s})$ and $p(s)$ are different functions.

The function $p(s)$ on $\mathscr{S}$ is such that

$$p(s) \geqslant 0 \quad \text{for all} \quad s \in \mathscr{S} \tag{3.3}$$

and

$$\Sigma_{\mathscr{S}} p(s) = 1 \tag{3.4}$$

**Definition**. Any function $p(s)$ on $\mathscr{S}$ satisfying (3.3) and (3.4) will be called an *unordered sampling design*. The function $p(s)$ may be induced by a given ordered design $p(\mathbf{s})$ on $\mathscr{S}^*$, or $p(s)$ may be postulated as a point of departure. ■

**Remark 1**. By our definition, a design is a *function* $p(\cdot)$, as described above. Some authors, for example, Godambe (1965), Hanurav (1962), refer to *the pair*, $(\mathscr{S}, p(\cdot))$, as the design. ■

**Example 1**. Let $\mathbf{s}_1, \ldots, \mathbf{s}_{15}$ be as in Example 3 of Section 1.2. Define $p(\mathbf{s})$ by $p(\mathbf{s}_i) = 0$ for $i = 1, 2, 3$ and $p(\mathbf{s}_i) = 1/12$ for $i = 4, 5, \ldots, 15$. The sets $A_{s_i}$ are $\{\mathbf{s}_1\}$ for $i = 1$; $\{\mathbf{s}_2\}$ $(i = 2)$; $\{\mathbf{s}_3\}$ $(i = 3)$; $\{\mathbf{s}_4, \mathbf{s}_5\}$ $(i = 4)$; $\{\mathbf{s}_6, \mathbf{s}_7\}$ $(i = 5)$; $\{\mathbf{s}_8, \mathbf{s}_9\}$ $(i = 6)$; $\{\mathbf{s}_{10}, \mathbf{s}_{11}, \ldots, \mathbf{s}_{15}\}$ $(i = 7)$. The induced unordered design is such that $p(s) = 0$ for $s = s_1, s_2, s_3$; $p(s) = 1/6$ for $s = s_4, s_5, s_6$; and $p(s) = 1/2$ for $s = s_7$. ■

The identifiability of units is a crucial factor which permits us to

(a) designate in any way we like a set of samples (sequences or sets) to which we intend to assign a positive probability of selection;
(b) distribute in any way we like the total probability mass among the members of this set.

Practical considerations will in various ways limit this freedom to choose. But in principle, (a) and (b) are features which distinguish sampling from finite populations from most techniques in traditional inference theory which is preoccupied with independent and identically distributed observations (even though it is often evident that data have not been gathered at random).

We shall find, under certain extensions of the basic model considered in Chapters 4–6, that "labels can be disregarded," so that the set of unlabeled measurements, $\{y_k : k \in s\}$, contains all the information relevant for the inference. There is also the possibility of deliberately "ignoring the labels," if this is deemed justifiable, for example, at the estimation stage; see Sections 2.4 and 3.5. Basically, however, one must take the position that since the labels are there, there is every reason to try to incorporate them in any inference-making process.

Any design, ordered or unordered, is characterized by a set of inclusion probabilities of first, second, and higher orders. In the sequence case, introduce two sets of sequences,

$$B_k = \{\mathbf{s} : k \in \mathbf{s}\} \quad (k = 1, \ldots, N)$$
$$B_{kl} = \{\mathbf{s} : k, l \in \mathbf{s}\} \quad (k \neq l = 1, \ldots, N)$$

For the set case, let

$$C_k = \{s : k \in s\} \quad (k = 1, \ldots, N)$$
$$C_{kl} = \{s : k, l \in s\} \quad (k \neq l = 1, \ldots, N)$$

**Definition.** *The inclusion probability*, $\alpha_k$, of unit $k$ is the probability of selecting that unit, that is, in the case of an ordered design $p(\mathbf{s})$,

$$\alpha_k = \Sigma_{B_k} p(\mathbf{s}) \quad (k = 1, \ldots, N)$$

The *second-order inclusion probability*, $\alpha_{kl}$, of units $k$ and $l$ is the probability of selecting both units $k$ and $l$, that is,

$$\alpha_{kl} = \Sigma_{B_{kl}} p(\mathbf{s}) \quad (k \neq l = 1, \ldots, N)$$

For an unordered design $p(s)$, the first- and second-order inclusion probabilities are, respectively,

$$\alpha_k = \Sigma_{C_k} p(s) \quad (k = 1, \ldots, N) \tag{3.5}$$
$$\alpha_{kl} = \Sigma_{C_{kl}} p(s) \quad (k \neq l = 1, \ldots, N)$$ ■

It is immediately clear how we should define higher-order inclusion probabilities, $\alpha_{k_1 k_2 \ldots k_n}$.

**Example 2.** In Example 1, $\alpha_1 = 10/12$, which is obtained from $\Sigma_{B_1} p(\mathbf{s})$ with $B_1 = \{\mathbf{s}_i : i \neq 2, 3, 8, 9\}$. Also, $\alpha_{23} = 8/12$ is obtained from $\Sigma_{B_{23}} p(\mathbf{s})$ where $B_{23} = \{\mathbf{s}_i : i = 8, 9, \ldots, 15\}$. The same results are obtained by summation of the induced probabilities $p(s)$ over the sets $C_1$ and $C_{23}$, respectively. ■

It is easy to see that $0 \leqslant \alpha_k \leqslant 1$ and that $0 \leqslant \alpha_{kl} \leqslant \min(\alpha_k, \alpha_l)$. The following theorem relates certain distribution properties of the effective sample size to the inclusion probabilities.

Define first the *expected effective sample size*, denoted $\nu$, as

$$\nu = E\{\nu(\mathbf{S})\} = \Sigma_{\mathscr{S}^*} \nu(\mathbf{s})\, p(\mathbf{s}) \tag{3.6}$$

and the *variance of the effective sample size* as

$$V\{\nu(\mathbf{S})\} = \Sigma_{\mathscr{S}^*} \{\nu(\mathbf{s}) - \nu\}^2\, p(\mathbf{s})$$

In the set case, the same definitions apply if $\mathbf{s}$ is replaced throughout by $s$, and $\mathscr{S}^*$ by $\mathscr{S}$.

In the following theorem given by Hanurav (1966), parts (i), (ii), and (iii) are due, respectively, to Godambe (1955), Hanurav (1962), and Yates and Grundy (1953):

**Theorem 1.1.** *Let $p$ be an arbitrary ordered design. Then the following holds:*

*(i)* $\Sigma_1^N \alpha_k = \nu$

*(ii)* $\Sigma\Sigma_{k\neq l}^{N}\alpha_{kl} = \nu(\nu - 1) + V\{\nu(\mathbf{S})\}$

*(iii)* *if* $\nu(\mathbf{s}) = \nu$ *for all* $\mathbf{s}$ *such that* $p(\mathbf{s}) > 0$, *then*

$$\Sigma\Sigma_{k\neq l}^{N}\alpha_{kl} = \nu(\nu - 1)$$

*and*

$$\Sigma_{k\neq l}^{N}\alpha_{kl} = (\nu - 1)\alpha_l$$

**Proof.** Let $\delta_k(\mathbf{S})$ be an indicator random variable taking the value 1 when $k \in \mathbf{S}$, and 0 otherwise. Statement (i) follows by noting that $E\{\delta_k(\mathbf{S})\} = \alpha_k$ and $\Sigma_1^N \delta_k(\mathbf{S}) = \nu(\mathbf{S})$:

$$\Sigma_1^N \alpha_k = \Sigma_1^N E\{\delta_k(\mathbf{S})\} = E\{\Sigma_1^N \delta_k(\mathbf{S})\} = E\{\nu(\mathbf{S})\} = \nu$$

In order to prove (ii), we also note that $\{\delta_k(\mathbf{s})\}^2 = \delta_k(\mathbf{s})$ for all $\mathbf{s}$, and that $E\{\delta_k(\mathbf{S})\,\delta_l(\mathbf{S})\} = \alpha_{kl}$ for $k \neq l$. Hence

$$\begin{aligned}\Sigma\Sigma_{k\neq l}^{N}\alpha_{kl} &= \Sigma\Sigma_{k\neq l}^{N} E\{\delta_k(\mathbf{S})\,\delta_l(\mathbf{S})\} = E\{\Sigma\Sigma_{k\neq l}^{N}\delta_k(\mathbf{S})\,\delta_l(\mathbf{S})\}\\ &= E[\{\Sigma_1^N \delta_k(\mathbf{S})\}^2 - \Sigma_1^N \{\delta_k(\mathbf{S})\}^2] = E[\{\nu(\mathbf{S})\}^2] - \nu\\ &= V\{\nu(\mathbf{S})\} + \nu(\nu - 1)\end{aligned}$$

Finally, statement (iii) follows by noting that if $\nu(\mathbf{s}) = \nu$ for all $\mathbf{s}$ such that $p(\mathbf{s}) > 0$, then $V\{\nu(\mathbf{S})\} = 0$, and we also have

$$\begin{aligned}\Sigma_{k\neq l}^{N}\alpha_{kl} &= \Sigma_{k\neq l}^{N} E\{\delta_k(\mathbf{S})\,\delta_l(\mathbf{S})\} = E\{\delta_l(\mathbf{S})\,\Sigma_{k\neq l}^{N}\delta_k(\mathbf{S})\}\\ &= E[\delta_l(\mathbf{S})\{\nu - \delta_l(\mathbf{S})\}] = (\nu - 1)\,\alpha_l\end{aligned}$$

□

**Remark 2.** Theorem 1.1 applies in the set case with obvious modifications. ■

## 1.4. SOME SPECIFIC SAMPLING DESIGNS

In this section we shall describe a number of designs that have proven important in survey sampling. We also prove a theorem dealing with the practical implementation, through a so-called sampling scheme, of a given sampling design. Finally, we discuss how to construct a sampling scheme such that a set of predetermined inclusion probabilities are attained. First some terminology is introduced.

**Definition.** A sampling design $p(\cdot)$ (ordered or unordered) is called a *noninformative design* if and only if $p(\cdot)$ is a function that does not depend on the $y$-values associated with the labels in $\mathbf{s}$ or $s$. ■

**Remark 1.** The requirement of noninformativeness of a design does not rule out the use of designs such that $p(\mathbf{s})$ or $p(s)$ is a function of the labeled auxiliary variable values $(k, x_k)$ $(k = 1, \ldots, N)$ as is the case with "probability-proportional-to-size" designs; see below. ■

One argument in favor of limiting consideration to noninformative designs derives from a practical advantage: The sampler can once and for all make the selection of the sample in his office, and then have his field workers collect the remaining part of the data $\{(k, y_k) : k \in s\}$. This way no recalculation of probabilities is needed.

Unless otherwise stated, all designs considered in the sequel are noninformative. Thus, for example, "any design $p(\cdot)$" should be taken to mean "any noninformative design $p(\cdot)$."

Examples of informative designs include situations where the sampler decides conditionally on already drawn labels and their associated $y$-values, what labels to draw next. Such sequential sampling plans have been considered by Basu (1969), Zacks (1969). Other examples of informative designs have been discussed by Lindley (1971a) ("length-biased" sampling), and by Scott (1975b).

Informative designs are of rather limited interest from a practical point of view, and we shall discuss only noninformative ones. This has some important consequences for the inferences arrived at, for example, in Chapters 5 and 6: Whether the inference be "classical" or "Bayesian," the choice of predictor is not influenced by the choice of design.

The following two definitions concern designs such that the sample size is fixed:

**Definition.** An ordered design $p(\mathbf{s})$ is called a *fixed size design*, abbreviated *FS design*, if $n(\mathbf{s})$ is constant for all $\mathbf{s} \in \mathscr{S}^*$ such that $p(\mathbf{s}) > 0$. If the constant size equals $n$, the abbreviation will be: *FS(n) design*.

A ordered design $p(\mathbf{s})$ (or an unordered design $p(s)$) is called a *fixed effective size design*, abbreviated *FES design*, if $\nu(\mathbf{s})$ ($\nu(s)$) is constant for all $\mathbf{s}$ ($s$) such that $p(\mathbf{s}) > 0$ ($p(s) > 0$). If the constant size equals $n$, the abbreviation will be: *FES(n) design*. ■

Hence, in our terminology, the FES property relates to a fixed number of *distinct* units, both in the sequence case and in the set case. Clearly, $n(\mathbf{s}) = n$ does not imply $\nu(\mathbf{s}) = n$; nor does $\nu(\mathbf{s}) = n$ imply $n(\mathbf{s}) = n$.

A design that is not an FS design will be called a non-FS design. Similarly, the term non-FES design will be used with obvious meaning. In Example 1 of Section 1.3, $p(\mathbf{s})$ is a non-FS design and also a non-FES design.

The *sampling fraction* will be denoted by $f_s = \nu(s)/N$. For an FES($n$) design, $f_s = f$ for all $s$ such that $p(s) > 0$, where $f = n/N$.

In a wide sense of the term we can say that a *with replacement* design is an ordered design such that the probability $p(\mathbf{s})$ is positive for some $\mathbf{s}$ in which at least one unit $k$ is repeated. A *without replacement* design is an ordered design such that any sequence containing at least one repeat has zero probability. Although the word "replacement" indicates a sequence of draws, any unordered design will also be considered of the without replacement variety, since repeats do not occur in a set $s$.

In simple random sampling procedures, any sequence $\mathbf{s}$ of fixed size $n(\mathbf{s}) = n$, or any set $s$ of fixed effective size $\nu(s) = n$, has the same positive probability.

In a design executed by probability proportional to size sampling (abbreviated PPS sampling), the probability of selecting a given sample is proportional to some function of the sample units other than the $y$-values; usually the size measure is related to auxiliary variable measurements $x_k$.

A number of common designs, that is, a number of specified *functions* $p(\cdot)$ on $\mathscr{S}^*$ or on $\mathscr{S}$, are tied to simple random sampling and PPS sampling. In order to denote such functions, we shall use lower case letter combinations, starting with *srs* for simple random sampling designs and with *pps* for PPS sampling designs. These letter combinations may require further qualification in order to define uniquely a design. Six often-used designs will now be defined, namely, *srsr, srs, ppsr, ppsrx, ppsux*, and *ppsx*:

(1) The symbol *srsr* will denote the design $p(\cdot)$ on $\mathscr{S}^*$ such that $p(\mathbf{s}) = 1/N^n$ for each sequence of size $n(\mathbf{s}) = n$, and $p(\mathbf{s}) = 0$ for all other sequences. The final *r* in *srsr* indicates the with replacement feature; this design is commonly known as simple random sampling with replacement (with a fixed number of draws). Clearly, *srsr* is a FS($n$) but non-FES design (unless $n = 1$); the inclusion probabilities are: $\alpha_k = 1 - (1 - 1/N)^n$ $(k = 1, \ldots, N)$.

(2) The symbol *srs* will denote the function $p(\cdot)$ such that $p(\mathbf{s}) = 1/N^{(n)}$ for every sequence consisting of $n(\mathbf{s}) = \nu(\mathbf{s}) = n$ components that are all different, where $N^{(n)} = N!/(N-n)!$. This design is commonly known as simple random sampling without replacement, hence the absence of an *r* after *srs*. Moreover *srs* will denote the induced set function, that is $p(s) = 1/\binom{N}{n}$ for each set $s$ containing exactly $n$ elements. Clearly, *srs* is an FS($n$) design (in the sequence case), and also an FES($n$)-design; the inclusion probabilities are $\alpha_k = f = n/N$ $(k = 1, \ldots, N)$.

(3) The symbol *ppsr* will denote the function $p(\mathbf{s})$ such that unit $k$ has probability $p_k$ of being selected in each of $n$ independent draws, where the outcome of the $i$th draw determines the $i$th component of $\mathbf{s}$. The $p_k$ are known positive numbers satisfying $\Sigma_1^N p_k = 1$. Thus *ppsr* is an FS($n$) and non-FES design with $\alpha_k = 1 - (1 - p_k)^n$.

(4) The symbol *ppsrx* will denote the particular case of *ppsr* such that $p_k$ is taken to be proportional to $x_k$, the known positive value for unit $k$ of the auxiliary variable $x$, that is, letting $\bar{x} = \Sigma_1^N x_k/N$,

$$p_k = \frac{x_k}{N\bar{x}} \quad (k = 1, \ldots, N) \tag{4.1}$$

There are various without replacement forms of PPS sampling:

(a) The successive sampling design, *ppsux*, defined below;
(b) Various *rejective sampling designs* (for examples, see the end of this section and Section 5.4);
(c) The *Rao–Hartley–Cochran (1962) design* (see Section 7.3).

(5) The symbol *ppsux* will denote the successive sampling design $p(\mathbf{s})$ such that in $n$ draws, one unit at a time, without replacement, the $j$th draw is performed with probability proportional to size of $x$ for the *remaining* units, for $j = 1, \ldots, n$. That is, let $p_k$ be given by (4.1) and draw label $k$ with probability $p_k$ in the first draw, and with probability

$$\frac{p_k}{1 - \Sigma_{i=1}^{j-1} p_{k_i}}$$

in the $j$th draw ($j = 2, 3, \ldots, n$) given that the units $k_1, \ldots, k_{j-1}$ were obtained in the first $j - 1$ draws, and $k \neq k_i$ ($i = 1, \ldots, j - 1$). Thus *ppsux* is an FS($n$) as well as an FES($n$) design.

(6) The symbol *ppsx* will be used to denote any FES($n$) design $p(s)$ such that the inclusion probabilities are $\alpha_k = np_k$ ($k = 1, \ldots, N$), where $p_k$ is given by (4.1). We assume that $x_k > 0$ and that $nx_k < N\bar{x}$ for all $k$. More than one design *ppsx* may exist for given values $x_1, \ldots, x_N$.

By a *sampling scheme* we shall mean a draw-by-draw mechanism for selecting units such that there is a predetermined set of selection probabilities for each unit in each draw. A sampling scheme is said to *implement* a given design $p(\mathbf{s})$ or $p(s)$ if the draw-by-draw mechanism reproduces the probabilities $p(\mathbf{s})$ or $p(s)$.

It is possible to find a sampling scheme to implement any given design, as shown in the sequence case by the following theorem due to T. V. H. Rao (1962). Additional discussion is found in Godambe (1965) and Hanurav (1966).

For a given design $p(\mathbf{s})$ let $n_0$ be the largest of the integers $n(\mathbf{s})$ for $\mathbf{s}$ such that $p(\mathbf{s}) > 0$. Extend any sequence $\mathbf{s}$ with $\nu(\mathbf{s}) < n_0$ by putting zeroes in the $n_0 - \nu(\mathbf{s})$ final positions. Thus all extended sequences $\mathbf{s}$ will have exactly $n_0$ components. A zero means that no unit at all is selected in the corresponding draw. Let $k = 0$ denote that no unit is drawn.

The sampling scheme will be characterized by $n_0$ sets of probabilities, the $i$th set being $p_i(k)$ ($k = 0, 1, \ldots, N$), where $p_i(k)$ is the probability of selecting unit $k$ at draw $i$, conditional on the labels that have been selected in draws $1, \ldots, i - 1$. Thus $p_i(0)$ is the probability of selecting no label at all in draw $i$, conditional on the result of previous draws. We must have, for each fixed $i$,

$$p_i(k) \geqslant 0 \quad (k = 0, 1, \ldots, N); \quad \Sigma_{k=0}^{N} p_i(k) = 1 \tag{4.2}$$

**Theorem 1.2.** *For any given design $p(\mathbf{s})$ there exists at least one sampling scheme that implements $p(\mathbf{s})$*

**Proof.** Consider extended sequences $\mathbf{s} = (k'_1, k'_2, \ldots, k'_{n_0})$. For fixed $k_1$, $k_2, \ldots$ (not necessarily distinct), set $A_1 = \{\mathbf{s} : k'_1 = k_1\}$; $A_2 = \{\mathbf{s} : k'_1 = k_1, k'_2 = k_2\}$, etc. In general, $A_i (i = 1, \ldots, n_0)$ is the set of extended sequences $\mathbf{s}$ having unit $k_j$ as its $j$th component for $j = 1, \ldots, i$. Set

$$p_1(k_1) = \Sigma_{A_1} p(\mathbf{s}) \tag{4.3}$$

and, for $i = 2, 3, \ldots, n_0$, conditional on $k_1, \ldots, k_{i-1}$ having been selected,

$$p_i(k_i) = \frac{\Sigma_{A_i} p(\mathbf{s})}{p_{i-1}(k_{i-1})} \tag{4.4}$$

provided the denominators are positive. Let $\mathbf{s} = (k_1, k_2, \ldots, k_{n_0})$ be any given extended sequence such that $p(\mathbf{s}) > 0$. Then $p_{i-1}(k_{i-1})$ in (4.4) is positive for $i = 2, \ldots, n_0$, and

$$p(\mathbf{s}) = \prod_{i=1}^{n_0} p_i(k_i)$$

that is, $p(\mathbf{s})$ is implemented by the sampling scheme defined by (4.3) and (4.4). □

A methodological problem arising, for example, in connection with use of the Horvitz–Thompson estimator and the generalized difference estimator (introduced in Section 1.6) is the following:

What would be a simple and practical way to implement a design on which the only requirement is that the first-order inclusion probabilities $\alpha_1, \ldots, \alpha_N$ have predetermined values? Consider the design *ppsx*. Given that $x_k > 0$, $nx_k < N\bar{x}$ for all $k$, the definition of *ppsx* requires only that $\alpha_k = nx_k/N\bar{x}$ for $k = 1, \ldots, N$. That is, all higher order inclusion probabilities are permitted to take arbitrary values. But for certain practical reasons (see below), additional requirements may be imposed, for example, that the second-order inclusion probabilities have certain properties.

In order to implement the design *ppsx*, we must find, for $i = 1, \ldots, n$, a set of probabilities $p_i(k)$ satisfying (4.2), and such that $\alpha_k \propto x_k$ $(k = 1, \ldots, N)$. Since *ppsx* is an FES($n$) design, $p_i(0) = 0$ for $i = 1, \ldots, n$.

The problem is not simple to resolve in a fashion which is appealing from a practical point of view. Various solutions exist and there is a considerable amount of literature on the topic, for example, Brewer (1963a), Carroll and Hartley (1964), Fellegi (1963), J. N. K. Rao (1963b, 1965), Durbin (1967), Hanurav (1962, 1967), Sampford (1967). Several of the suggested methods, of which reviews are found in Rao and Bayless (1969), Bayless and Rao (1970), Sampford (1975) are operational only in the case $n = 2$.

If a non-FES design were acceptable, a simple method yielding the desired set of $\alpha_k$ would be to conduct a Bernoulli trial for each unit $k$, with probability of success equal to $\alpha_k$. If the trial for unit $k$ results in success, this unit is included in the sample, otherwise not. Under this design, the effective sample size is a random variable $\nu(S)$ with expected value $\nu = E\{\nu(S)\} = \Sigma_1^N \alpha_k$.

If an FES($n$) design is required, there is still a fairly simple method to achieve the desired $\alpha_k$ ($k = 1, \ldots, N$). The method, as first given by Madow (1949), works as follows:

Let $Q$ be a random variable with uniform distribution over the [0, 1] interval. Let $g_k = \Sigma_1^k \alpha_j$, $k = 1, \ldots, N$, and $g_0 = 0$. The procedure is defined by selecting unit $k$ if $Q$ takes the value $q$ such that

$$g_{k-1} \leqslant i - 1 + q < g_k$$

for some $i = 1, \ldots, n$. Letting $a_{ki} = g_{k-1} - i + 1$, $b_{ki} = g_k - i + 1$, it is easily seen that the procedure gives the required $\alpha_k$:

$$\Sigma_{i=1}^n P(a_{ki} \leqslant Q < b_{ki}) = \alpha_k \quad (k = 1, \ldots, N)$$

In this method, the sample is essentially a systematic sample, which makes many of the $\alpha_{kl}$ equal to zero. This is the prime drawback of the otherwise simple method, because a set of nonzero $\alpha_{kl}$ is needed, for example, in obtaining an estimate of the variance of the Horvitz–Thompson estimator; see Remark 2 of Section 7.5. Goodman and Kish's (1950) modification of the method removes this drawback. The resulting $\alpha_{kl}$ are, however, difficult to calculate.

Several of the methods devised to achieve a given set of $\alpha_k$ are given in terms of probabilities $p_i(k)$ that change from one draw $i$ to the next, thus requiring a considerable amount of recalculation. The methods of Brewer (1963a), Fellegi (1963), Durbin (1967), and Hanurav (1967) are of this kind.

One of the three methods of Sampford (1967) is particularly appealing. The probabilities are constant from draw to draw (excepting the first draw). On the other hand, the rejective sampling feature of the method may be seen as a drawback: Units are drawn one by one with replacement, using in the first draw the probabilities $p_1(k) = \alpha_k/n$, and in each of the subsequent $n - 1$ draws $p_i(k) \propto \alpha_k/(1 - \alpha_k)$ ($k = 1, \ldots, N$; $i = 2, \ldots, n$). Any sample not containing $n$ distinct labels is rejected, that is, if a label is repeated, the procedure starts from scratch and continues until $n$ distinct labels have been selected. The derivation of the $\alpha_{kl}$ for this method again requires considerable computational effort; good approximations of the $\alpha_{kl}$ have, however, been given by Asok and Sukhatme (1976).

## 1.5. DATA AND ESTIMATORS

The interest in the foundations of survey sampling was sparked by the realization that a finite population, because of the identifiability of its units, admits classes of estimators fundamentally different from those customarily dealt with in traditional statistical inference.

We give first a simple example of such an "unconventional" estimator. Also in this section we define the concept of the data available in the survey sampling inference problem. Further, we discuss the general concept of an estimator of the finite population mean.

**Example 1.** Contrary to what one might think, the sample mean $\bar{y}_S$ is not the uniformly minimum variance estimator under simple random sampling, as shown in this example due to Royall (1968).

Let $N = 3$, $n = 2$, and $s_1 = \{1, 2\}$, $s_2 = \{1, 3\}$, $s_3 = \{2, 3\}$. Under the design *srs*, we have $p(s_i) = 1/3$ for $i = 1, 2, 3$.

Define the estimator $t$ by

$$t = \begin{cases} t_1 = y_1/2 + y_2/2 & \text{if} \quad s_1 \quad \text{occurs} \\ t_2 = y_1/2 + 2y_3/3 & \text{if} \quad s_2 \quad \text{occurs} \\ t_3 = y_2/2 + y_3/3 & \text{if} \quad s_3 \quad \text{occurs} \end{cases}$$

Under the given design, $t$ is unbiased for the population mean, since $\Sigma_1^3 t_i p(s_i) = (y_1 + y_2 + y_3)/3 = \bar{y}$. The variance of $t$ (see Section 1.6) is $V(t) = \Sigma_1^3 (t_i - \bar{y})^2 p(s_i)$. By comparison, the variance of the sample mean $\bar{y}_S = \Sigma_S y_k/2$ is $V(\bar{y}_S) = \Sigma_1^3 (\bar{y}_{Si} - \bar{y})^2 p(s_i)$.

Computation shows that, under the design *srs*,

$$V(\bar{y}_S) - V(t) = \frac{y_3(3y_2 - 3y_1 - y_3)}{54}$$

Hence $V(t) < V(\bar{y}_S)$ for all parameter vectors $\mathbf{y} = (y_1, y_2, y_3) \in R_3$ such that $y_3(3y_2 - 3y_1 - y_3) > 0$, and $\bar{y}_S$ is not a uniformly minimum variance estimator.

The estimator $t$ would never be considered in traditional statistical inference, since $t$ depends on the labels in $s$. ■

The example raises the question: If an estimator is construed as a function of the data obtained from the sampling, what exactly do we mean by *the data* in survey sampling?

Once an ordered sample $\mathbf{s}$ has been selected and its member units have been measured, then the result can be specified as the sequence of pairs $(k_i, y_{k_i})$ for $i = 1, \ldots, n(\mathbf{s})$. In the case of an unordered sample $s$, the result can be specified as the set of pairs $(k, y_k)$ for $k \in s$.

Some frequently used estimators (see Section 1.6) require additional information in order to be computed, such as the values $x_k$ of an auxiliary variable $x$, or the probabilities $\alpha_k$ with which the various units $k$ are selected, when sampling is with unequal probabilities.

We shall assume that such information is available in a complete list of the population. Thus, once a sample **s** or $s$ has been selected, we can go back and consult the list to obtain the value $x_k$ or $\alpha_k$ for each of the sampled units.

**Definition.** By *the data*, **d**, obtained from the observation of a sequence **s** and the associated $y$-values, we mean the sequence of pairs

$$\mathbf{d} = ((k_1, y_{k_1}), \ldots, (k_{n(\mathbf{s})}, y_{k_{n(\mathbf{s})}}))$$

which will be written more compactly as

$$\mathbf{d} = ((k, y_k)\colon k \in \mathbf{s}) \tag{5.1}$$

(As for the notation $k \in \mathbf{s}$, see Remark 3 of Section 1.2.)

By the data, $d$, obtained from the observation of a set $s$ and the associated $y$-values, we mean the set of pairs

$$d = \{(k, y_k)\colon k \in s\} \qquad \blacksquare \tag{5.2}$$

Sometimes we shall be interested in the values $y_k$ only, and not in the label part of the data.

**Definition.** In the sequence case, the *unlabeled data*, $\mathbf{y_s}$, obtained from **d** is defined as the vector

$$\mathbf{y_s} = (y_{k_1}, \ldots, y_{k_{n(\mathbf{s})}})$$

or, for short,

$$\mathbf{y_s} = (y_k\colon k \in \mathbf{s})$$

In the set case, the unlabeled data, $y_s$, obtained from $d$, is the set

$$y_s = \{y_k\colon k \in s\} \qquad \blacksquare$$

**Remark 1.** A comment is required about our proposed usage of the set notation $y_s$. We shall assume that if, for example, two units in the sample $s$ have identical $y_k$-values, this value will appear twice in $y_s$. Thus if $d = \{(1, 100), (3, 110), (4, 100\}$, then the unlabeled data are $y_s = \{100, 100, 110\}$, which is not to be identified with the set $\{100, 110\}$, as under strict adherence to practice in set theory. Thus, for any given $s$, each of the two sets, $s$ and $y_s$, contains $\nu(s)$ elements, but those of $y_s$ are not necessarily distinct. $\blacksquare$

**Example 2.** In the sequence case we can reconstruct **d** from the information **s** and $\mathbf{y_s}$. By contrast, in the set case, it is not possible to reconstruct $d$ from the information $s$ and $y_s$.

Suppose that $\mathbf{s} = (4, 6, 1, 6)$ and that $y_1 = y_4 = 60$, $y_6 = 50$. Hence $\mathbf{y_s} = (60, 50, 60, 50)$. Since the ordering is the same in **s** as in $\mathbf{y_s}$, we conclude that $\mathbf{d} = ((4, 60), (6, 50), (1, 60), (6, 50))$. Knowledge of $\mathbf{s}$ distinguishes the two 60-valued components of $\mathbf{y_s}$.

Reducing $\mathbf{s}$ to $s = r(\mathbf{s})$, we obtain $s = \{1, 4, 6\}$, $d = \{(1, 60), (4, 60), (6, 50)\}$ and $y_s = \{50, 60, 60\}$. Clearly, $d$ can not be reconstructed from $s$ and $y_s$. ■

Next we define the concepts of sample space and statistic.

**Definition.** The sample space of the random variable $\mathbf{D}$ taking values $\mathbf{d}$ given by (5.1) is

$$\mathscr{X}^* = \{\mathbf{d}: \mathbf{s} \in \mathscr{S}^*, \mathbf{y} \in \Omega\} \tag{5.3}$$

The sample space of the random variable $D$ taking values $d$ given by (5.2) is

$$\mathscr{X} = \{d: s \in \mathscr{S}, \mathbf{y} \in \Omega\} \quad ■ \tag{5.4}$$

**Remark 2.** We have called $\mathbf{s}$ (or $s$) a sample, $\mathbf{d}$ (or $d$) the data, and now $\mathscr{X}^*$ (or $\mathscr{X}$) is introduced as the sample space. A more complete description might have been "sample space of the data." ■

**Definition.** In the sequence case, a *statistic*, $Z = u(\mathbf{D})$, is a (not necessarily real-valued) function on $\mathscr{X}^*$ such that, for any given $\mathbf{s} \in \mathscr{S}^*$, $u(\cdot)$ depends on $\mathbf{y}$ only through those $y_k$ for which $k \in \mathbf{s}$.

In the set case, a statistic $Z = u(D)$ is a function on $\mathscr{X}$ such that, for any given $s \in \mathscr{S}$, $u(\cdot)$ depends on $\mathbf{y}$ only on those $y_k$ for which $k \in s$. ■

We observed in Section 1.2 that there exists a function $r(\mathbf{s})$ which reduces the sequence $\mathbf{s}$ into the set $s = r(\mathbf{s})$ by dismissing information about order and multiplicity of units. Similarly any observed data $\mathbf{d}$ can be reduced into $d = u_0(\mathbf{d})$, say, by dismissing the same information. The reduction function $u_0(\cdot)$ defines a statistic $D = u_0(\mathbf{D})$ on $\mathscr{X}^*$ which will be of particular interest in Chapter 2, where the probability distribution of $\mathbf{D}$ and of $D$ will also be discussed.

For any two ordered samples, $\mathbf{s}_1$ and $\mathbf{s}_2$, with corresponding data vectors $\mathbf{d}_1$ and $\mathbf{d}_2$, respectively, it follows that $r(\mathbf{s}_1) = r(\mathbf{s}_2)$ if and only if $u_0(\mathbf{d}_1) = u_0(\mathbf{d}_2)$.

**Example 3.** Let $\mathbf{s}_1 = (2, 8, 3, 2)$ and $\mathbf{d}_1 = ((2, 40), (8, 40), (3, 25), (2, 40))$. Further, let $\mathbf{s}_2 = (3, 2, 8, 8)$ and $\mathbf{d}_2 = ((3, 25), (2, 40), (8, 40), (8, 40))$. Then $r(\mathbf{s}_1) = r(\mathbf{s}_2) = \{2, 3, 8\}$ and $u_0(\mathbf{d}_1) = u_0(\mathbf{d}_2) = \{(2, 40), (3, 25), (8, 40)\}$. ■

Among the various statistics that can be defined on the sample spaces $\mathscr{X}^*$ and $\mathscr{X}$, we are primarily interested in those that are suitable as *estimators* of $\bar{y} = \Sigma_1^N y_k/N$. The general notation for an estimator will be $t(\mathbf{D})$ or $t(D)$. Frequently, we shall simply write $t$ in place of $t(\mathbf{D})$ or $t(D)$.

In survey sampling we have, as already pointed out, the possibility to choose any convenient design $p(\cdot)$, thereby determining the distribution of any given estimator $t$.

But for the moment, consider the design $p(\cdot)$ to be given and consider only the choice of estimator $t$. Minimization of variance or mean square error has traditionally been the predominating concern in survey sampling theory. Historically, theorists in sampling as well as in traditional statistical theory have proceeded by defining a suitably large class of estimators and by finding the best (for example, in the sense of minimum variance) estimator in that class.

In survey sampling, Hansen and Hurwitz (1943) were among the first to present an unbiased estimator based on unequal probability sampling. Their estimator, see formula (6.8), can be described as "label-dependent." A label-dependent estimator (see Remark 1 of Section 1.6) is one for which knowledge of the labels, and not only of the unlabeled data $y_s$, is necessary for computation of the value that the estimator takes for a given sample. Narain (1951) is another early reference on unequal probability sampling.

The idea of various classes of linear estimators for finite populations was given more explicit definition in the influential paper of Horvitz and Thompson (1952). For the first time it was made clear how identifiability of units is instrumental in creating new classes of linear estimators.

Koop (1963) carried this idea one step further in designing a system of classes of linear estimators.

A complete discussion of the various linear classes is not essential for our presentation. Further references in this respect are Godambe (1955), Godambe and Joshi (1965), Hanurav (1966), Basu (1971). Starting with the papers by Godambe and Joshi (1965) and Hanurav (1966), emphasis was being shifted into consideration of classes of general (linear or nonlinear) estimators.

A linear estimator that disregards the identifiability of units would be of the type

$$t = \Sigma_{i=1}^{n} c_i y_{K_i}$$

Here, the coefficients are indexed only by the drawing order $i$. The estimator $t$ depends on the data $\mathbf{d}$ only through the unlabeled data $\mathbf{y_s}$. This is the kind of estimator typically considered in traditional statistical theory.

But, given the identifiability, the finite population setup permits certain generalizations:

**Definition.** By a *linear estimator, t*, we shall mean an estimator such that

$$t = t(\mathbf{D}) = w_{0\mathbf{S}} + \Sigma_{\mathbf{S}} w_{k\mathbf{S}} y_k \tag{5.5}$$

for the sequence case, and

$$t = t(D) = w_{0S} + \Sigma_S w_{kS} y_k \tag{5.6}$$

for the set case, where $w_{0\mathbf{S}}, w_{k\mathbf{S}}$ $(w_{0S}, w_{kS})$ are constants not depending on the $y$-values, but possibly on auxiliary information. ■

**Definition.** An estimator is said to be a *linear homogeneous estimator* if and only if it is of the form

$$t = t(\mathbf{D}) = \Sigma_{\mathbf{S}} w_{k\mathbf{S}} y_k \tag{5.7}$$

in the sequence case and

$$t = t(D) = \Sigma_S w_{kS} y_k \tag{5.8}$$

in the set case. ■

**Example 4.** The estimator $t$ in Example 1 of Section 1.2 is of the form (5.8) with $w_{ks} = 1/6\ p(s)$ for all $k \in s$. The estimator $t$ in Example 1 of this section is of the form (5.8) with $w_{1s_1} = w_{1s_2} = w_{2s_1} = w_{2s_3} = 1/2; w_{3s_2} = 2/3$; $w_{3s_3} = 1/3$. ■

The class of *linear* estimators consists of all estimators of the form (5.5) (or (5.6) in the set case), that is this class comprises all homogeneous and nonhomogeneous linear estimators. The class of *linear homogeneous* estimators consists of all estimators of the form (5.7) (or (5.8) in the set case). In the class of *all* estimators the restriction to linearity is removed.

**Remark 3.** Basu (1971) pointed out that the sample space $\mathscr{X}$ defined by (5.4) is not linear. A linear estimator such as (5.6), while linear in $y_k$ for $k \in S$, is not linear in the data $D$. Basu (1971) proposed the name *generalized linear* estimator for an estimator of the type (5.6). ■

## 1.6. SPECIFIC ESTIMATORS AND UNBIASEDNESS

We now introduce notation and exact definition of a series of estimators of $\bar{y}$ well known in survey sampling. At this point they serve mainly as examples of estimators that survey samplers have found useful and interesting. The motivation behind each estimator will have to be saved until later chapters. The various designs to be mentioned were introduced in Section 1.4. Also discussed in this section are the concepts of design unbiasedness, variance, and mean square error.

The *sample mean* is defined by

$$\bar{y}_{\mathbf{S}} = t(\mathbf{D}) = \frac{\Sigma_{\mathbf{S}}\, y_k}{n(\mathbf{S})} \tag{6.1}$$

In particular, if the design is an FS($n$) design,

$$\bar{y}_{\mathbf{S}} = t(\mathbf{D}) = \frac{\Sigma_{\mathbf{S}}\, y_k}{n} \tag{6.2}$$

The *sample mean of distinct units* is defined by

$$\bar{y}_S = t(D) = \frac{\Sigma_S\, y_k}{\nu(S)} \tag{6.3}$$

In particular, under an FES($n$) design,

$$\bar{y}_S = t(D) = \frac{\Sigma_S\, y_k}{n} \tag{6.4}$$

In many practical situations we have access to a known vector $\mathbf{x} = (x_1, \ldots, x_N)$ of values of an auxiliary variable $x$. Many well-known estimators capitalize on this information.

The *ratio estimator* (or *ratio-of-the-means estimator*) is defined by

$$t_R = t(D) = \frac{\bar{x}\bar{y}_S}{\bar{x}_S} \tag{6.5}$$

in the set case, where $\bar{y}_S$ is given by (6.3), $\bar{x} = \Sigma_1^N\, x_k/N$, $\bar{x}_S = \Sigma_S x_k/\nu(S)$, $x_k$ being the auxiliary variable measurement for unit $k$. In the sequence case, an analogous ratio estimator is arrived at by replacing $S$ by $\mathbf{S}$ and $\nu(S)$ by $n(\mathbf{S})$.

Introduce next the following *mean-of-the-ratios* statistics useful when positive auxiliary values $x_1, \ldots, x_N$ are known:

In the sequence case, let

$$R_{yx}^{\circ} = \frac{\Sigma_{\mathbf{S}}\, y_k/x_k}{n(\mathbf{S})} \tag{6.6}$$

and in the set case, let

$$R_{yx} = \frac{\Sigma_S\, y_k/x_k}{\nu(S)} \tag{6.7}$$

The *Hansen-Hurwitz (1943) estimator* is defined for the design *ppsr* as

$$t_{HH} = t(\mathbf{D}) = \Sigma_{\mathbf{S}} \frac{y_k}{Np_k} \tag{6.8}$$

In particular, for the design *ppsrx*, the Hansen–Hurwitz estimator becomes

$$t_{HH} = t(\mathbf{D}) = \bar{x}R_{yx}^{\circ} \tag{6.9}$$

where $R_{yx}^{\circ}$ is given by (6.6) with $n(\mathbf{S}) = n$, since *ppsrx* is an FS($n$) design.

The *Horvitz–Thompson (1952) estimator* is defined for an arbitrary design as

$$t_{HT} = t(D) = \Sigma_S \frac{y_k}{N\alpha_k} \tag{6.10}$$

where $\alpha_k$ is the inclusion probability of unit $k$ ($k = 1, \ldots, N$). In particular, for the design *ppsx*, the Horvitz–Thompson estimator becomes

$$t_{HT} = t(D) = \bar{x} R_{yx} \tag{6.11}$$

where $R_{yx}$ is given by (6.7) with $\nu(s) = n$, since *ppsx* is an FES($n$) design.

The *Raj (1956) estimator* is defined for the design *ppsux* as

$$t_{DR} = t(\mathbf{D}) = \frac{\Sigma_{i=1}^{n} t_i}{nN} \tag{6.12}$$

where $t_1 = y_{K_1}/p_{K_1}$ and, for $i = 2, 3, \ldots, n$,

$$t_i = \Sigma_{j=1}^{i-1} y_{K_j} + \frac{(1 - \Sigma_{j=1}^{i-1} p_{K_j}) y_{K_i}}{p_{K_i}}$$

The *difference estimator* is defined as

$$t_D = t(D) = \bar{y}_S + c(\bar{x} - \bar{x}_S) \tag{6.13}$$

where $c$ is a constant. An analogous definition is possible in the sequence case by replacing $S$ by $\mathbf{S}$ and $D$ by $\mathbf{D}$.

The *generalized difference estimator*, suggested by Basu (1971) as a modified form of $t_{HT}$, is defined for an arbitrary design as

$$t_{GD} = t(D) = \Sigma_S \frac{y_k - e_k}{N\alpha_k} + \bar{e} \tag{6.14}$$

where $e_k$ is an arbitrary but known number attached to unit $k (k = 1, \ldots, N)$, $\bar{e} = \Sigma_1^N e_k/N$, and $\alpha_k$ is the inclusion probability of unit $k$.

In particular, if $e_k = cx_k$ for $k = 1, \ldots, N$, where $c$ is a predetermined constant and $x_k$ is the known value for unit $k$ of an auxiliary variable, then $t_{GD}$ takes the form

$$t_{GD} = \Sigma_S \frac{y_k}{N\alpha_k} + c\left(\bar{x} - \Sigma_S \frac{x_k}{N\alpha_k}\right) \tag{6.15}$$

Specific choices of design bring us back to estimators already introduced:

For the design *ppsx*, the special form (6.15) of $t_{GD}$ becomes identical to the special form (6.11) of $t_{HT}$.

For the design *srs*, $t_{GD}$ given by (6.15) becomes identical to the difference estimator $t_D$ given by (6.13).

Each of the introduced estimators has merit under specific conditions, as will be seen in later chapters. At this point we shall observe how the estimators relate to the concept of design unbiasedness now to be introduced.

**Remark 1.** An estimator can be considered "label-dependent" or "label-independent." The meaning of these terms is made more precise by the following discussion which refers to the set case.

The data $D$ can be reduced to varying extents by throwing away part of the information in $D$. For example, the pair $(S, y_S)$ is a reduction of $D$, because $D$ can not be reconstructed from knowledge of $(S, y_S)$ alone. Moreover, the unlabeled data $y_S$ is an even further reduction of $D$. Knowledge of $y_S$ only does not permit reconstruction of $(S, y_S)$, let alone of $D$.

An estimator is said to be *label-independent* if and only if it can be constructed without knowledge of the set $S$ of labels that corresponds to the data $D$.

An estimator that can be constructed from $D$, or from a reduction of $D$, is said to be *label-dependent* as long as $S$ has not been reduced away. We consider that $S$ has been "reduced away" whenever the reduction of $D$ under consideration does not permit reconstruction of $S$.

For example, $\bar{y}_S$ is label-independent. It can be constructed from knowledge of the unlabeled data $y_S$ alone, and $y_S$ does not contain information about $S$. In fact, an even further reduction of $D$, namely, $(\nu(S), \Sigma_S y_k)$ would suffice to compute $\bar{y}_S$.

Examples of label-dependent estimators are $t_{HT}$ and $t_{GD}$. Whether they be written on the form (6.10), (6.11), (6.14), or (6.15), we must know nothing less than $D$ in order to compute these estimators. It is assumed that, for each $k \in s$, we can go to the complete list of the population to obtain the required value $\alpha_k$, $e_k$ or $x_k$.

However, the classical sampling theory estimators $t_D$ and $t_R$ are also label-dependent. We can certainly compute $t_D$ and $t_R$ if all we know about $D$ is the reduction $(S, y_S)$. Since the information $S$ is retained, we can go to the list of the population to obtain the values $x_k$ necessary for computation of the required quantity $\Sigma_S x_k$. Note that an even further reduction, namely $(S, \Sigma_S y_k)$, will actually suffice in order to compute $t_D$ and $t_R$. ■

In the Fixed population approach considered in Chapters 2 and 3, the only randomness derives from the man-made randomness imposed in form of the design, $p(\mathbf{s})$ or $p(s)$. The distribution of $\mathbf{D}$ or $D$, or any function $t(\cdot)$ of $\mathbf{D}$ or $D$, is entirely determined by the chosen design and the state of the parameter $\mathbf{y}$. The following definitions relate to properties of interest for the distribution of an estimator $t(\cdot)$ under a given design: unbiasedness, variance, mean square error, and so on. Note that these concepts are being defined with reference to *a specific design*. For example the variance of a given estimator $t$ may take totally different values for two different designs. The definitions below are given only in the set case; their counterparts for the sequence case are completely analogous.

**Definition.** An estimator $t = t(D)$ is said to be *p-unbiased*, or *design unbiased*, if and only if

$$E(t) = \Sigma_{\mathscr{S}} t(d)p(s) = \bar{y} \quad \text{for all} \quad \mathbf{y} \in \Omega$$

The quantity

$$B(t) = E(t) - \bar{y}$$

is called the *p-bias*, or *design bias*, of $t = t(D)$.

The *p-variance* of an estimator $t = t(D)$ is defined as

$$V(t) = E\{t - E(t)\}^2 = \Sigma_{\mathscr{S}}\{t(d) - E(t)\}^2\, p(s)$$

The *p-MSE* (where MSE stands for Mean Square Error) of an estimator $t = t(D)$ is defined by

$$\text{MSE}(t) = E(t - \bar{y})^2 = \Sigma_{\mathscr{S}}\{t(d) - \bar{y}\}^2\, p(s)$$

If the $p$-bias $B(t)$ is zero, then $\text{MSE}(t) = V(t)$ ■

Since the $p$-variance is determined in part by the choice of design, it is evident that a goal such as the attainment of minimal variance must sometimes be approached through a twofold choice procedure, involving the choice of *a pair* $(p, t)$, where $p$ is a design and $t$ is an estimator.

**Definition.** A pair $(p, t)$ will be called *a strategy*. A strategy is said to be *p-unbiased* if the estimator $t$ is $p$-unbiased when used with the design $p$. The *p-variance* and the *p-MSE* of a strategy $(p, t)$, denoted $V(p, t)$ and $\text{MSE}(p, t)$, are defined respectively as the $p$-variance and the $p$-MSE of the estimator $t$ under the design $p$. ■

The following examples illustrate the concepts just defined.

**Example 1.** As is easily checked, the Horvitz–Thompson estimator (6.10) as well as the two forms (6.14) and (6.15) of the generalized difference estimator are $p$-unbiased if the design is such that $\alpha_k > 0$ $(k = 1, \ldots, N)$.

The Hansen–Hurwitz estimators (6.8) and (6.9) and the Raj estimator (6.12) are $p$-unbiased under the designs *ppsr*, *ppsrx*, and *ppsux*, respectively.

Under the design *srsr*, the sample means (6.2) and (6.3) are $p$-unbiased. Moreover, under the design *srs*, the form (6.4) of $\bar{y}_S$ and the difference estimator (6.13) are $p$-unbiased.

The ratio estimator (6.5), finally, is $p$-biased under the design *srs*. In classical sampling theory, the ratio estimator always was very popular, due among other things to the fact that its $p$-bias under *srs* is small, of order $n^{-1}$; see Cochran (1963). Later it was discovered that there is indeed a design for which $t_R$ is $p$-unbiased, namely the unequal probability sampling design associated with the names of Hájek (1949) and Lahiri (1951); see the discussion in Section 7.3. Refinements of the technique were given by Midzuno (1952) and Sen (1953). This work on the ratio estimator was important in that it directed interest towards unequal probability sampling in general. ■

**Example 2.** From the definition of $p$-unbiasedness it follows that a homogeneous linear estimator of the form (5.8) is $p$-unbiased if and only if $\Sigma_{\mathscr{S}} \Sigma_s w_{ks} y_k p(s) = \bar{y}$ for all $\mathbf{y} \in \Omega$, that is, if and only if the set of $N$ conditions,

$$\Sigma_{C_k} w_{ks} p(s) = \frac{1}{N} \quad (k = 1, \ldots, N) \tag{6.16}$$

hold, where $C_k = \{s : k \in s\}$. The Horvitz–Thompson estimator $t_{HT}$ given by (6.10) is of the form (5.8) with $w_{ks} = 1/N\alpha_k$ for $k \in s$. Using the identity (3.5), we see that these coefficients satisfy (6.16). ■

**Example 3.** The Horvitz-Thompson estimator (6.10) is unique in the following sense: $t_{HT}$ is the only $p$-unbiased homogeneous linear estimator of the type $\Sigma_s w_k y_k$. In this type of estimator the weight of $y_k$ may change with the label $k$, but, for any given $k$, the weight stays the same for all $s$. The uniqueness claim is easily established by setting $w_{ks} = w_k$ in (6.16). We get, using (3.5), $w_k \alpha_k = 1/N$ $(k = 1, \ldots, N)$, which is satisfied only by $t_{HT}$. ■

**Example 4.** A general linear estimator of the type (5.6) is $p$-unbiased if and only if the $N + 1$ conditions,

$$\Sigma_S w_{0s} p(s) = 0$$

$$\Sigma_{C_k} w_{ks} p(s) = \frac{1}{N} \quad (k = 1, \ldots, N)$$

are simultaneously satisfied. It is easily checked that the conditions are satisfied by the generalized difference estimator $t_{GD}$ for which $w_{0s} = \bar{e} - \Sigma_s e_k/N\alpha_k$ and $w_{ks} = 1/N\alpha_k$. ■

## 1.7. SOME REFLECTIONS ON THE BASIC MODEL

A mathematical model serves as a formalization of certain features of the real world context surrounding a given scientific problem. Often the *purpose* of a model is simply to clarify our understanding of the problem at hand. A model that it successful in this respect may still be a less than perfect representation of the real world; that is, the model may be more or less *realistic*.

We have reasoned that the model introduced in this chapter provides, more adequately than earlier attempts, a suitable framework for inference in survey sampling.

Recent literature has shown that the model has served the purpose of clarifying and making more rigorous the discussion of the foundations of survey sampling. The discussion has incorporated most of the established sampling procedures worked out especially in the 1930s and 1940s. The basic model has helped in relating these procedures to standard principles of inference worked out independently by researchers in traditional statistical theory.

The basic model manages to preserve many features of the survey sampling reality, including the identifiability of the units of the finite population. Many populations dealt with in practice are not only finite but also made up of identifiable units. (Yet the Fisherian framework in traditional statistical inference has so far not lent itself very well to representations beyond the random sampling variety.) But the basic model for survey sampling adopted above is not perfect either as a representation of the survey statistician's reality. Future work in the area will no doubt bring about refinements and extensions.

Our final comments in this introductory chapter relate to the limitations of the basic model as well as to the limitations of this book:

(i) We assume that units possess some identifying characteristic which makes possible listing of units and provides opportunity to contact a given unit in order to obtain a measurement of the target variable. Also, it is assumed that the population size $N$ is known.

These assumptions mean that certain finite populations are not within the framework of this book. Examples are populations of type "the fish in a lake" (Hanurav, 1966), where a list of the units is lacking or impossible to construct, and where possibility to contact units may be nonexistant.

When available, the "identifying characteristic" is transformed into a numerical value in the form of the label $k$ ($k = 1, \ldots, N$). In a sense, the label constitutes a very basic kind of auxiliary variable measurement. Whether such auxiliary information is relevant or not will have to be decided from case to case. Cochran (1963) and Dalenius (1957) give examples where labels can be used as auxiliary information.

(ii) We assume that the process of listing the population can be undertaken without error. In reality, the listed population may differ from the target population: Some units that should have been included are actually excluded, and vice versa. Also by error, the same unit may become listed several times, through two or more separate labels.

(iii) The assumption made in Section 1.2–1.4 of absence of measurement error is sometimes unrealistic. For example, in the case of a nonresponse from unit $k$, a measurement $y_k$ is totally lacking. Even where a measurement has been obtained, it may not be "true," but affected by systematic error and/or response error of random nature. Such situations can probably be treated through extensions of the basic model; one attempt in this direction is Koop (1974).

(iv) In practice, a survey would seldom be undertaken with the purpose of investigating only one variable of interest; usually there are many. However, in the absence of a unified theory for the estimation of many means, our principal concern in this book must be with the single mean situation.

(v) Inference about the *population mean* is only one aspect, but an important one, of finite population inference. Other parametric functions of interest are, for example, the mode, various population quantiles, the population variance. Also, it is frequently desired to make inference for subpopulations on the basis of a sample drawn from the total population; such questions are left unanswered in the present text.

## 1.8. LIST OF NOTATION

For easy future reference we give below a list of notation introduced so far. Sections 4.2 and 4.3 will introduce additional notation necessary to deal with the Superpopulation approach of Chapters 4–7.

**a. Population**

| | |
|---|---|
| $\mathcal{U} = \{1, 2, \ldots, N\}$ | Finite population |
| $N$ | Population size |
| $\mathbf{y} = (y_1, \ldots, y_N)$ | Parameter vector |
| $\bar{y} = \Sigma_1^N y_k/N$ | Population mean of variable of interest |
| $\bar{x} = \Sigma_1^N x_k/N$ | Population mean of auxiliary variable |
| $\Omega$ | Parameter space |

**b. Sample = Ordered sample = Sequence**

| | |
|---|---|
| $\mathbf{s} = (k_1, \ldots, k_{n(\mathbf{s})}) = (k : k \in \mathbf{s})$ | Sequence of not necessarily distinct units |
| $n(\mathbf{s})$ | Sample size = number of components of $\mathbf{s}$ |
| $\nu(\mathbf{s})$ | Effective sample size = number of distinct components of $\mathbf{s}$ |
| $k \in \mathbf{s}$ | (Somewhat unconventional) notation for "$k$ is a component of the sequence $\mathbf{s}$" |
| $\mathbf{d} = ((k, y_k): k \in \mathbf{s})$ | Labeled data = sequence of pairs $(k_i, y_{k_i})$, where $k_i (i = 1, \ldots, n(\mathbf{s}))$ are the components of $\mathbf{s}$ |
| $\mathbf{y_s} = (y_k: k \in \mathbf{s})$ | Unlabeled data = the sequence of $y$-values in $\mathbf{d}$ taken in the order of the elements of $\mathbf{s}$ |

**c. Sample = Unordered sample = Reduced sample = Set**

| | |
|---|---|
| $s = \{k : k \text{ is a component of } \mathbf{s}\}$ | Set of distinct units in $\mathbf{s}$, or sample of distinct units without order |
| $r(\cdot)$ | "Reducing function" |
| $s = r(\mathbf{s})$ | The set $s$ obtained from reducing $\mathbf{s}$ |

| | |
|---|---|
| $A_s = \{\mathbf{s} : r(\mathbf{s}) = s\}$ | Set of sequences $\mathbf{s}$ that reduce to the same set $s$ |
| $\nu(s)$ | Effective sample size of $s$ (note that $\nu(\mathbf{s}) = \nu\{r(\mathbf{s})\} = \nu(\mathbf{s})$) |
| $d = \{(k, y_k) : k \in s\}$ | Labeled data = set of pairs $(k, y_k)$,where $k \in s$ |
| $y_s = \{y_k : k \in s\}$ | Unlabeled data = the set of $y$-values in $d$ |
| $f_s = \nu(s)/N$ | Sampling fraction |
| $f = n/N$ | Sampling fraction for an FES($n$) design |

**d. Ordered sampling design**

| | |
|---|---|
| $\mathscr{S}^* = \{\mathbf{s}\}$ | Set of all ordered samples $\mathbf{s}$ |
| $p = p(\mathbf{s})$ | Ordered sampling design = real function on $\mathscr{S}^*$ such that $0 \leqslant p(\mathbf{s}) \leqslant 1$ and $\Sigma_{\mathscr{S}^*} p(\mathbf{s}) = 1$ |
| $B_k = \{\mathbf{s} : k \in \mathbf{s}\}$ | Set of all sequences $\mathbf{s}$ of which $k$ is a component |
| $\alpha_k = \Sigma_{B_k} p(\mathbf{s})$ | Inclusion probability of unit $k$ |

**e. Unordered sampling design**

| | |
|---|---|
| $\mathscr{S} = \{s\}$ | Set of all unordered samples $s$ |
| $p = p(s)$ | Unordered sampling design = real function on $\mathscr{S}$ such that $0 \leqslant p(s) \leqslant 1$ and $\Sigma_{\mathscr{S}} p(s) = 1$ |
| $C_k = \{s : k \in s\}$ | The set of all sets $s$ of which $k$ is a member |
| $\alpha_k = \Sigma_{C_k} p(s)$ | Inclusion probability of unit $k$ |

**f. Particular designs**

| | |
|---|---|
| noninformative design | $p(\mathbf{s})$ ($p(s)$) does not explicitly depend on the $y$-values |
| FS($n$) design | Fixed sample size design: $p(\mathbf{s}) > 0 \Rightarrow n(\mathbf{s}) = n$ |
| FES($n$) design | Fixed effective sample size design: $p(\mathbf{s}) > 0 \Rightarrow \nu(\mathbf{s}) = n$ (equivalently for $s$) |
| *srsr* | Design of simple random sampling with replacement; $p(\mathbf{s}) = 1/N^n$ for all sequences $\mathbf{s}$ containing $n$ components |
| *srs* | Design of simple random sampling without replacement; $p(s) = 1/\binom{N}{n}$ for all sets $s$ containing $n$ elements |
| *ppsr* | The PPS-sampling design with replacement such that unit $k$ is drawn with given probability $p_k$ in each of $n$ draws |
| *ppsrx* | The design *ppsr* with $p_k = x_k/\Sigma_1^N x_k$ |
| *ppsux* | The PPS without replacement design such that unit $k$ is drawn with PPS of $x$ of remaining units (successive sampling) |
| *ppsx* | Any FES($n$) design with $\alpha_k = nx_k/\Sigma_1^N x_k$ |

**g. Random variables**

| | |
|---|---|
| $\mathbf{S}$ | Random sequence taking values $\mathbf{s} \in \mathscr{S}^*$ |
| $p(\mathbf{s}) = P(\mathbf{S} = \mathbf{s})$ | Distribution of $\mathbf{S}$; ordered design |
| $K_i$ | Random variable taking values $k_i$, $k_i = 1, \ldots, N$, where $k_i$ is the $i$th component of $\mathbf{S}$ |
| $\mathbf{D} = ((k, y_k)\colon k \in \mathbf{S})$ | Data in the sequence case |
| $\mathscr{X}^* = \{\mathbf{d}\colon \mathbf{s} \in \mathscr{S}^*, \mathbf{y} \in \Omega\}$ | Sample space in the sequence case |
| $S$ | Random set taking values $s \in \mathscr{S}$ |
| $p(s) = P(S = s)$ | Distribution of $S$; unordered design |
| $D = \{(k, y_k)\colon k \in S\}$ | Data in the set case |
| $\mathscr{X} = \{d\colon s \in \mathscr{S}, \mathbf{y} \in \Omega\}$ | Sample space in the set case |

**h. Estimators**

| | |
|---|---|
| $t(\mathbf{D})$, $t(D)$, or simply $t$ | Estimator; a real function on $\mathscr{X}^*$ or on $\mathscr{X}$ |
| $t = w_{0S} + \Sigma_S w_{kS} y_k$ | Linear estimator (in the set case) |
| $t = \Sigma_S w_{kS} y_k$ | Homogeneous linear estimator (in the set case) |
| $\bar{y}_\mathbf{S} = \Sigma_\mathbf{S} y_k / n(\mathbf{S})$ | Sample mean |
| $\bar{y}_S = \Sigma_S y_k / \nu(S)$ | Sample mean of distinct units |
| $t_R = \bar{x}\bar{y}_S / \bar{x}_S$ | Ratio estimator |
| $t_{HH} = \Sigma_\mathbf{S}\, y_k / nNp_k$ | Hansen–Hurwitz estimator for the design *ppsr* |
| $t_{DR} = \Sigma_1^n\, t_i / nN$ | Raj estimator for the design *ppsux* |
| $t_{HT} = \Sigma_S y_k / N\alpha_k$ | Horvitz–Thompson estimator |
| $t_D = \bar{y}_S + c(\bar{x} - \bar{x}_S)$ | Difference estimator |
| $t_{GD} = \Sigma_S (y_k - e_k)/N\alpha_k + \bar{e}$ | Generalized difference estimator |

**i. Expectation operators**

| | |
|---|---|
| $E(t) = \Sigma_{\mathscr{S}}\, p(s) t(d)$ | $p$-Expectation of $t = t(D)$ |
| $V(t) = E\{t - E(t)\}^2$ | $p$-Variance of $t = t(D)$ |
| $\mathrm{MSE}(t) = E(t - \bar{y})^2$ | $p$-Mean square error of $t = t(D)$ |
| $B(t) = E(t) - \bar{y}$ | $p$-Bias of $t = t(D)$ |

CHAPTER 2

# Inference under the Fixed Population Model: The Concepts of Sufficiency and Likelihood

In Chapter 1 we distinguished two different outlooks on survey sampling theory, the Fixed population approach and the Superpopulation approach. Chapters 2 and 3 deal exclusively with the former approach, that is, it is assumed that with each unit $k$ is associated a fixed but unknown real number $y_k$. We assume also that if unit $k$ belongs to the sample, then $y_k$ can be observed without error. Thus nonsampling errors are absent.

In the Fixed population approach, the *only* stochastic element upon which an inference can be based is the one introduced through the sampling design $p$. The terms $p$-unbiased, $p$-variance, and $p$-MSE were defined in Section 1.6. In their place, we shall, in Chapters 2 and 3, simply use the terms unbiased, variance, and MSE, respectively, since there is no source of ambiguity. (In Chapters 4 and following, it becomes, however, essential to distinguish, for example, $p$-unbiasedness from $\xi$-unbiasedness.)

Chapters 2 and 3 present some of the more important results in non-Bayesian inference under the Fixed population approach. Chapter 2 is devoted to the role of the sufficiency and likelihood concepts in survey sampling. In Chapter 3 we shall discuss criteria for choosing and judging estimators and strategies, including admissibility, uniformly minimum variance unbiasedness, minimaxity, and invariance. The order of the sections does not necessarily follow the historical development of the topic.

In Chapter 2 in particular, the reader is reminded to carefully observe the distinction between the notation $\mathbf{S}, \mathbf{s}, \mathbf{D}, \mathbf{d}$, and $\mathbf{y_s}$ pertaining to the sequence case and the notation $S, s, D, d$, and $y_s$ pertaining to the set case, as well as the distinction between $\bar{y}_{\mathbf{S}}$ and $\bar{y}_S$; see the list of symbols in Section 1.8.

## 2.1. THE INFERENCE PROBLEM

In traditional statistical theory, an inference problem is often described in the following terms: Let $X$ be a random quantity to be observed. Let $\mathscr{X}$ be the sample space of $X$, and let $\mathscr{A}$ be a $\sigma$-algebra of subsets of $\mathscr{X}$. Let $\mathscr{P} = \{P_\theta : \theta \in \Omega\}$ be a family of probability measures $P_\theta$ on $\mathscr{A}$, such that each probability measure is indexed by a parameter $\theta$ belonging to a specified parameter space $\Omega$. We observe $X$ without knowing the true value of $\theta \in \Omega$, and the observation is to serve as a basis for inference about $\theta$.

In the present chapter we shall consider the problem of estimating the finite population mean $\bar{y} = \Sigma_1^N y_k/N$, on the basis of an observation on the random quantity $\mathbf{D}$ or $D$ introduced in Section 1.5. As pointed out by Basu (1969) this problem can be seen to have the structure of the general inference problem described above.

**Remark 1.** The random quantity observed will sometimes be taken to be $\mathbf{D}$, sometimes $D$. Under an unordered sampling design, we observe $D$. Under an ordered sampling design, we may observe $\mathbf{D}$, which in this case is the most complete description of the outcome. But even under an ordered design, we shall often, without loss of information, record only $D$, thus deliberately neglecting information about order and multiplicity of units; the justification for this procedure is given in Section 2.2. ■

In Theorem 2.1 below, expressions are given for the probability that the random quantity $\mathbf{D}$ (or $D$) will take any specified value $\mathbf{d} = ((k, y_k) : k \in \mathbf{s})$ (or $d = \{(k, y_k) : k \in s\}$, respectively). In order to express these probabilities in a compact manner we need the following definition.

**Definition.** A specified value $\mathbf{d} = ((k, y_k) : k \in \mathbf{s})$ is said to be *consistent* with a particular population vector $\mathbf{y}_0 = (y_{01}, \ldots, y_{0N})$ if and only if $y_k = y_{0k}$ for all $k \in \mathbf{s}$. A specified value $d = \{(k, y_k) : k \in s\}$ is said to be *consistent* with $\mathbf{y}_0$ if and only if $y_k = y_{0k}$ for all $k \in s$. ■

In other words, a value $\mathbf{d}$ (or $d$) is consistent with a particular population vector if and only if the $y$-values of the units in $\mathbf{s}$ (or $s$), as given by $\mathbf{d}$ (or $d$), coincide with the $y$-values of the same units as given by the population vector. (See also Example 1.)

**Theorem 2.1.** *For any given ordered design p and for any population vector* $\mathbf{y}$, *the probability that the random quantity* $\mathbf{D}$ *will take a value* $\mathbf{d} = ((k, y_k) : k \in \mathbf{s})$ *is given by*

$$p_{\mathbf{y}}(\mathrm{d}) = \begin{cases} p(\mathrm{s}) & \text{if } \mathbf{d} \text{ is consistent with } \mathbf{y} \\ 0 & \text{otherwise} \end{cases} \tag{1.1}$$

*For any design p and for any population vector* $\mathbf{y}$, *the probability that the random quantity D will take a value* $d = \{(k, y_k) : k \in s\}$ *is given by*

$$p_{\mathbf{y}}(d) = \begin{cases} p(s) & \text{if } d \text{ is consistent with } \mathbf{y} \\ 0 & \text{otherwise} \end{cases} \tag{1.2}$$

**Proof.** The expression (1.1) is easily derived, since

$$\begin{aligned} P(\mathbf{D} = \mathbf{d}) &= P(\mathbf{D} = \mathbf{d} \wedge \mathbf{S} = \mathbf{s}) \\ &= P(\mathbf{S} = \mathbf{s})P(\mathbf{D} = \mathbf{d} \mid \mathbf{S} = \mathbf{s}) \end{aligned}$$

and the conditional probability $P(D = \mathbf{d} \mid \mathbf{S} = \mathbf{s})$ equals unity if $\mathbf{d}$ is consistent with $\mathbf{y}$, and 0 otherwise. The expression (1.2) is derived in the same manner. □

**Example 1.** Let $N = 3$, and suppose that $\mathbf{y} = (a, b, c)$. Then the outcome $d = \{(1, a), (2, b)\}$ is consistent with $\mathbf{y}$, and has probability $p_{\mathbf{y}}(d) = p(s)$, where $s = \{1, 2\}$. The outcome $d = \{(1, b), (2, a)\}$ is not consistent with $\mathbf{y}$ (provided $a \neq b$) and has zero probability. ■

When $\mathbf{D}$ is observed, the sample space $\mathscr{X}^*$ of $\mathbf{D}$ is defined by formula (5.3) of Section 1.5. Let the $\sigma$-algebra $\mathscr{A}$ be the set of all subsets of $\mathscr{X}^*$. For any given design $p$ and for any vector $\mathbf{y}$ we have a discrete probability measure on $\mathscr{A}$, specified by the function (1.1). This probability measure depends on $\mathbf{y}$, which acts as the unknown parameter vector. We assume that, before any sampling is done, it is known that $\mathbf{y}$ must belong to some specified set $\Omega$, which is our parameter space. Following the prevalent practice in current literature we assume in Chapters 2 and 3 that $\Omega = R_N$, unless otherwise stated.

**Remark 2.** For each $\mathbf{s} \in \mathscr{S}^*$, exactly one outcome $\mathbf{d}$ is consistent with a given vector $\mathbf{y}$. Hence, for a given $\mathbf{y}$, the number of outcomes $\mathbf{d}$ with positive probability is equal to the number of samples $\mathbf{s}$ with positive probability, which is at most countable. For any $A \in \mathscr{A}$ the probability measure $P_{\mathbf{y}}(A)$ is given by the sum $\Sigma p_{\mathbf{y}}(\mathbf{d})$ over those $\mathbf{d} \in A$ for which $p_{\mathbf{y}}(\mathbf{d}) > 0$. ■

**Remark 3.** Recall that the sample space, as we have defined it, is the set of all possible outcomes $\mathbf{d}$ of $\mathbf{D}$, *not* the set $\mathscr{S}^*$ of all possible samples. ■

When we consider $D$, instead of $\mathbf{D}$, the sample space $\mathscr{X}$ is defined by formula (5.4) of Section 1.5. The $\sigma$-algebra $\mathscr{A}$ is the set of all subsets of $\mathscr{X}$. For any given design $p$ and for any vector $\mathbf{y}$ we have a discrete probability measure on $\mathscr{A}$, specified by the function (1.2). Here, too, $\mathbf{y}$ is the unknown parameter vector, belonging to a parameter space $\Omega$, which equals $R_N$ unless otherwise stated. Remarks 2 and 3 above apply in this case also, after obvious modifications.

Let us pinpoint a few characteristics of the inference problem under consideration:

(i) There is a family of discrete probability measures.

(ii) The set of outcomes **d**(or $d$) having positive probability varies with the parameter vector **y**.

(iii) The number $N$ of unknown parameters is usually much larger than the number of units observed. The objective, however is *not* to estimate the parameters $y_1, \ldots, y_N$ *individually*, but only the parametric function $\bar{y} = \Sigma_1^N y_k/N$.

(iv) When **D** (or $D$) has been observed, we have exact knowledge of some of the parameters, namely those $y_k$ for which $k$ belongs to the sample.

(v) The probability distribution of **D** (or $D$) is determined by the unknown parameter vector **y** and by the sampling design chosen by the surveyor. Since the surveyor can affect the probability distribution of what he is going to observe, there is a need in survey sampling theory to distinguish different types of problem, namely,

(a) Choosing an estimator for a given design
(b) Choosing a design for a given estimator
(c) Choosing a strategy, that is, choosing simultaneously a combination of design and estimator.

Most of the results in Chapters 2 and 3 refer to situation (a), the most thoroughly researched of the three problems.

## 2.2. SUFFICIENCY

The sufficiency criterion is used, in survey sampling theory as in traditional statistical theory, as a means of finding a reduction of the observed data into a more compact form without loss of information essential to the inference problem at hand. Thus the search for estimators is often restricted to the class of estimators that are functions of a sufficient statistic. In this context the so-called Rao-Blackwell theorem is helpful (see Section 2.3).

The concept of sufficiency and the Rao-Blackwell theorem were first invoked in survey sampling by Basu (1958). Certain other sufficiency concepts, different from the conventional one, have also been considered in survey sampling. Linear sufficiency and distribution-free sufficiency are discussed by Godambe (1966), and Bayesian sufficiency by Godambe (1968). Linear sufficiency will be briefly considered at the end of this section. Distribution-free and Bayesian sufficiency refer to the Superpopulation approach; the latter concept will be considered in Chapter 6.

By "sufficiency" we mean in this section "conventional sufficiency." The main results to be shown are: If we delete from **D** the information about order and multiplicity of units, then the resulting statistic $D = u_0(\mathbf{D})$ is sufficient for **y**, and even minimal sufficient. The first rigorous proof of these facts was given by Basu and Ghosh (1967) and Basu (1969). Prior to that, however, the results appear to have been taken for granted by several authors.

Throughout this section we consider only ordered sampling designs. Let $p$ be any given ordered sampling design. The concept of a statistic was introduced in Section 1.5. The definition of *sufficiency* of a statistic is entirely in line with the one used in traditional statistical theory.

**Definition.** A statistic $Z = u(\mathbf{D})$ is said to be *sufficient* for the parameter vector $\mathbf{y}$ if and only if the conditional distribution of $\mathbf{D}$, given $Z = z$, does not depend on $\mathbf{y}$, provided this conditional probability is well defined. ■

Thus, a statistic $Z$ is sufficient for $\mathbf{y}$ if and only if, for any fixed $\mathbf{d}$ and $z$, the conditional probability $P(\mathbf{D} = \mathbf{d} \mid Z = z; \mathbf{y})$ is constant for all $\mathbf{y}$ such that $P(Z = z; \mathbf{y}) > 0$.

We recall from Section 1.2 that any ordered sample $\mathbf{s}$ can be reduced, by dismissing information about order and multiplicity of units, to an unordered sample $s = r(\mathbf{s})$, where $s$ is the set of all distinct units in the sequence $\mathbf{s}$. By the same dismissal of information, the observed data $\mathbf{d}$ can be reduced to $d = u_0(\mathbf{d})$. Using the function $u_0(\cdot)$, we define the statistic $D = u_0(\mathbf{D})$ such that, if $\mathbf{D}$ takes the value $\mathbf{d}$, then $D$ takes the value $u_0(\mathbf{d})$. This statistic is of special interest because, as shown in the next two theorems, it is *sufficient* for $\mathbf{y}$, and even *minimal sufficient*. Though this may appear to be an "intuitively obvious" fact, the proof is not trivial.

**Theorem 2.2.** *For any given ordered design $p$, $D = u_0(\mathbf{D})$ is a sufficient statistic for* $\mathbf{y}$.

**Proof.** Consider the conditional probability $P(\mathbf{D} = \mathbf{d} \mid D = d; \mathbf{y})$ for any fixed points $\mathbf{d}$ and $d$ in the sample spaces of $\mathbf{D}$ and $D$, respectively (Note: Here $d$ is *not* necessarily equal to the point $u_0(\mathbf{d})$.) This conditional probability is defined only for $\mathbf{y}$ such that $P(D = d; \mathbf{y}) > 0$, and is then given by

$$P(\mathbf{D} = \mathbf{d} \mid D = d; \mathbf{y}) = \frac{P(\mathbf{D} = \mathbf{d} \wedge D = d; \mathbf{y})}{P(D = d; \mathbf{y})} \tag{2.1}$$

We distinguish two cases (a) and (b) below.

(a) In the case where $d = u_0(\mathbf{d})$ we have from Theorem 2.1 that

$$P(\mathbf{D} = \mathbf{d} \wedge D = d; \mathbf{y}) = P(\mathbf{D} = \mathbf{d}; \mathbf{y}) = p_{\mathbf{y}}(\mathbf{d}) = p(\mathbf{s}) \tag{2.2}$$

since, under the assumptions made, $\mathbf{d}$ is consistent with $\mathbf{y}$. Moreover,

$$P(D = d; \mathbf{y}) = \sum_{\mathbf{s}' \in A_s} p(\mathbf{s}') \tag{2.3}$$

where $A_s = \{\mathbf{s}' : r(\mathbf{s}') = s\}$. (Note that $u_0(\mathbf{d}') = d$ if and only if $\mathbf{s}' \in A_s$.)

(b) In the case where $d \neq u_0(\mathbf{d})$ we have

$$P(\mathbf{D} = \mathbf{d} \wedge D = d; \mathbf{y}) = 0 \tag{2.4}$$

Summing up, we have from (2.1)–(2.4) that

$$P(\mathbf{D}=\mathbf{d} \mid D=d; \mathbf{y}) = \begin{cases} p(\mathbf{s}) / \sum_{\mathbf{s}' \in A_s} p(\mathbf{s}') & \text{for } d = u_0(\mathbf{d}) \\ 0 & \text{for } d \neq u_0(\mathbf{d}) \end{cases}$$

for all $\mathbf{y}$ such that $P(D=d; \mathbf{y}) > 0$. This conditional probability does not depend on $\mathbf{y}$. Hence $D$ is sufficient for $\mathbf{y}$. □

By a *partition* of some basic set $A$ we shall mean a collection of mutually disjoint subsets of $A$, called *partition sets*, the union of which is the whole set $A$. With any statistic $Z = u(\mathbf{D})$ we can associate a partition $\mathscr{P}_u$ of the sample space of $\mathbf{D}$, such that two outcomes, $\mathbf{d}_1$ and $\mathbf{d}_2$, belong to the same partition set if and only if $u(\mathbf{d}_1) = u(\mathbf{d}_2)$. Now we can define the concept of a minimal sufficient statistic.

**Definition.** A statistic $Z_1 = u_1(\mathbf{D})$ is said to be *minimal sufficient* for $\mathbf{y}$ if and only if, for any sufficient (for $\mathbf{y}$) statistic $Z = u(\mathbf{D})$, each partition set of $\mathscr{P}_u$ is a subset of a partition set of $\mathscr{P}_{u_1}$. ■

A minimal sufficient statistic may be interpreted as the strongest possible reduction of the original data that is still sufficient. The next theorem says that $D = u_0(\mathbf{D})$ is a minimal sufficient statistic.

**Theorem 2.3.** *For any given ordered design $p$, $D = u_0(\mathbf{D})$ is a minimal sufficient statistic for* $\mathbf{y}$.

**Proof.** We introduce the indicator function

$$\delta(\mathbf{d}, \mathbf{y}) = \begin{cases} 1 & \text{when } \mathbf{d} \text{ is consistent with } \mathbf{y} \\ 0 & \text{otherwise} \end{cases}$$

Now the probability function (1.1) from Theorem 2.1 can be expressed as

$$p_{\mathbf{y}}(\mathbf{d}) = p(\mathbf{s})\delta(\mathbf{d}, \mathbf{y}) \tag{2.5}$$

From the definition of sufficiency it follows that for any sufficient statistic $Z = u(\mathbf{D})$ we can write, provided that $P\{Z = u(\mathbf{d}); \mathbf{y}\} > 0$,

$$P\{\mathbf{D} = \mathbf{d} \mid Z = u(\mathbf{d}); \mathbf{y}\} = h(\mathbf{d}) \tag{2.6}$$

where $h(\cdot)$ does not depend on $\mathbf{y}$. Without imposing any restrictions we can assume that $h(\mathbf{d}) > 0$, and hence $p(\mathbf{s}) > 0$. (For, if $h(\mathbf{d}) = 0$, then $p_{\mathbf{y}}(\mathbf{d}) = 0$ for all $\mathbf{y} \in R_N$, and such outcomes may be deleted.)

From (2.6) it follows that

$$p_{\mathbf{y}}(\mathbf{d}) = P\{\mathbf{D} = \mathbf{d} \wedge Z = u(\mathbf{d}); \mathbf{y}\} = P\{Z = u(\mathbf{d}); \mathbf{y}\}h(\mathbf{d}) = g\{u(\mathbf{d}), \mathbf{y}\}h(\mathbf{d}) \tag{2.7}$$

where $g(\cdot,\cdot)$ depends on $\mathbf{d}$ only through $u(\mathbf{d})$.

Now we are prepared to prove the theorem. Let $\mathcal{P}_{u_0}$ be the partition defined by $D = u_0(\mathbf{D})$, and let $\mathcal{P}_u$ be the partition defined by an arbitrary sufficient statistic $Z = u(\mathbf{D})$. We shall prove that any partition set of $\mathcal{P}_u$ is a subset of a partition set of $\mathcal{P}_{u_0}$.

Let $\mathbf{d}_1$ and $\mathbf{d}_2$ be any two outcomes such that $u(\mathbf{d}_1) = u(\mathbf{d}_2)$. Thus $\mathbf{d}_1$ and $\mathbf{d}_2$ belong to the same partition set of $\mathcal{P}_u$. We shall see that $u(\mathbf{d}_1) = u(\mathbf{d}_2)$ implies that $u_0(\mathbf{d}_1) = u_0(\mathbf{d}_2)$, and hence $\mathbf{d}_1$ and $\mathbf{d}_2$ belong to the same partition set of $\mathcal{P}_{u_0}$.

Since $u(\mathbf{d}_1) = u(\mathbf{d}_2)$ we have from (2.7) that

$$p_{\mathbf{y}}(\mathbf{d}_1) = g\{u(\mathbf{d}_1), \mathbf{y}\}h(\mathbf{d}_1)$$

and

$$p_{\mathbf{y}}(\mathbf{d}_2) = g\{u(\mathbf{d}_2), \mathbf{y}\}h(\mathbf{d}_2) = g\{u(\mathbf{d}_1), \mathbf{y}\}h(\mathbf{d}_2)$$

Hence,

$$p_{\mathbf{y}}(\mathbf{d}_2) = \frac{h(\mathbf{d}_2)}{h(\mathbf{d}_1)}\, p_{\mathbf{y}}(\mathbf{d}_1) \tag{2.8}$$

and, from (2.5) and (2.8),

$$p(s_2)\delta(\mathbf{d}_2, \mathbf{y}) = \frac{h(\mathbf{d}_2)}{h(\mathbf{d}_1)}\, p(s_1)\delta(\mathbf{d}_1, \mathbf{y}) \tag{2.9}$$

From (2.9) we conclude that $\delta(\mathbf{d}_1, \mathbf{y}) = 1$ if and only if $\delta(\mathbf{d}_2, \mathbf{y}) = 1$. Hence $r(s_1) = r(s_2)$, which implies that $u_0(\mathbf{d}_1) = u_0(\mathbf{d}_2)$; that is, $\mathbf{d}_1$ and $\mathbf{d}_2$ belong to the same partition set of $\mathcal{P}_{u_0}$.

We have seen that if any two outcomes $\mathbf{d}_1$ and $\mathbf{d}_2$ belong to the same partition set, say $B$, of $\mathcal{P}_u$, then they also belong to the same partition set, say $B_0$, of $\mathcal{P}_{u_0}$. Thus $B \subseteq B_0$, and hence $D = u_0(\mathbf{D})$ conforms to the definition of a minimal sufficient statistic. □

**Remark 1.** In concluding this section we define and discuss the concept of *linear sufficiency*, which was applied to the survey sampling situation by Godambe (1966). The concept, which was introduced in the Gauss–Markov setup by Barnard (1963), has seldom been used in survey sampling, and we mention it primarily to indicate a variation of the traditional sufficiency theme.

Let

$$\mathbf{c}\mathbf{y}' = \Sigma_1^N c_k y_k$$

be a linear parametric function, where $\mathbf{c} = (c_1, \ldots, c_N)$ is a row vector, and $\mathbf{y}'$ is the transpose of the row vector $\mathbf{y}$. Let $t = \Sigma_S w_{kS} y_y$ be a linear estimator of $\mathbf{c}\mathbf{y}'$.

If, for each $s \in \mathscr{S}$, we define a vector

$$\mathbf{w}(s) = (w_{1s}, w_{2s}, \ldots, w_{Ns})$$

where $w_{ks} = 0$ as soon as $k \notin s$, we can write

$$t = \mathbf{w}(S)\mathbf{y}'$$

Now we can define linear sufficiency: A linear homogeneous estimator $t_0 = \mathbf{w}_0(S)\mathbf{y}'$ is said to be *linear sufficient* for the linear parametric function $\mathbf{cy}'$ if and only if, for any other linear homogeneous estimator $t = \mathbf{w}(S)\mathbf{y}'$, and for any $s \in \mathscr{S}$,

$$\mathbf{w}(s)\mathbf{w}_0(s)' = 0 \quad \Leftrightarrow \quad \mathbf{w}(s)\mathbf{c}' = 0$$

The population mean is a linear parametric function with $c_k = 1/N$ ($k = 1, \ldots, N$). Godambe (1966) proved that, for any given design, a linear homogeneous estimator $t = \mathbf{w}(S)\mathbf{y}'$ is linear sufficient for $\bar{y}$ if and only if it can be written as $t = g(S)\Sigma_S y_k$ (that is, $w_{ks} = g(s)$ for $k \in s$, and $w_{ks} = 0$ for $k \notin s$).

From Godambe's result it follows, for example, that the sample mean of the distinct units $\bar{y}_S = \Sigma_S y_k / \nu(S)$ is linear sufficient for $\bar{y}$, for any given design. ■

## 2.3. RAO–BLACKWELLIZATION

We shall now see how the sufficient statistic $D = u_0(\mathbf{D})$ may be utilized for estimation purposes, under an ordered design. The main result, which is an adaptation to survey sampling of the so-called Rao–Blackwell theorem, states that, for any estimator that depends on $\mathbf{D}$, we can find a better estimator (in the MSE sense) which depends on $\mathbf{D}$ only through $D = u_0(\mathbf{D})$. Moreover, the theorem gives a method for constructing such an improved estimator. We shall also see that, although the possibility exists of improving certain estimators, it is still not possible to find a unique "best" estimator. The discussion of this fact revolves around the "incompleteness" of $D$.

**Remark 1.** For a given design $p$, an estimator $t_1$ is said to be *at least as good* as another estimator $t_2$ if $\text{MSE}(t_1) \leqslant \text{MSE}(t_2)$ for all $\mathbf{y} \in R_N$. If, in addition, strict inequality holds for at least one $\mathbf{y}$, then $t_1$ is said to be *better* than $t_2$. ■

The Rao–Blackwell theorem was first considered in survey sampling by Basu (1958). Some results in the same direction were obtained earlier by Murthy (1957) and Raj and Khamis (1958) without explicit reference to the Rao–Blackwell theorem.

Let $p$ be any given ordered design, and let $t = t(\mathbf{D})$ be an estimator, not necessarily unbiased, of $\bar{y}$. For each possible outcome $d$ of $D = u_0(\mathbf{D})$ we define

$$t^*(d) = E\{t(\mathbf{D}) \mid D = d\} \tag{3.1}$$

Since $D$ is sufficient for **y**, the function $t^*(\cdot)$ does not depend on any unknown parameters and thus defines a real-valued statistic $t^* = t^*(D)$ that depends on **D** only through $D = u_0(\mathbf{D})$. The following theorem says that $t^*$ has some favorable properties, when compared to the original estimator $t$. The theorem is a direct adaptation to survey sampling of the well-known Rao–Blackwell theorem.

**Theorem 2.4.** *Let $t = t(\mathbf{D})$ be a (not necessarily unbiased) estimator of $\bar{y}$ and let $t^* = t^*(D)$ be defined by (3.1). Then*

*(a)* $E(t) = E(t^*)$,
*(b)* $\text{MSE}(t) = \text{MSE}(t^*) + E\{(t - t^*)^2\}$,
*(c)* $\text{MSE}(t^*) \leqslant \text{MSE}(t)$

*with strict inequality in (c) for all $\mathbf{y} \in R_N$ such that $P(t \neq t^*; \mathbf{y}) > 0$.*

**Proof.** (a) Using conditional expectation we see that

$$E(t) = E\{E(t \mid D)\} = E(t^*)$$

(b) We obtain

$$\text{MSE}(t) = E(t - \bar{y})^2 = E\{(t - t^*) + (t^* - \bar{y})\}^2 = E(t - t^*)^2 + \text{MSE}(t^*)$$

since the product term vanishes.

Finally, (c) follows immediately from (b). □

The estimator $t^*$ is often said to be constructed by the process of "Rao–Blackwellization" of the original estimator $t$. Since $t^*$ depends on **D** only through $D$, it is independent of the order and multiplicity of units in the sample **S**. In survey sampling, the Rao–Blackwell theorem thus implies that for any estimator $t$ that depends on the order and multiplicity of units in **S**, we can find a better estimator (as measured by the MSE) that does not depend on order and multiplicity. For this reason we can from here on limit discussion (with some exceptions, mainly in Chapter 7) to estimators that do not depend on order and multiplicity of units.

We shall illustrate the Rao–Blackwellization technique by a few examples.

**Example 1.** A sequence of $n$ units is drawn by the design *srsr*. An unbiased estimator of $\bar{y}$ is the sample mean of the $n$ units drawn, possible repeats included, $t = \bar{y}_\mathbf{S} = \Sigma_\mathbf{S} y_k/n$, with variance $V(\bar{y}_\mathbf{S}) = \sigma^2/n$, where $\sigma^2 = \Sigma_1^N (y_k - \bar{y})^2/N$. Every unit enters in $\bar{y}_\mathbf{S}$ as many times as it appears in the sequence **S**. Thus $\bar{y}_\mathbf{S}$ depends on the multiplicity of units, but not on the order. The Rao–Blackwell technique gives

$$t^*(d) = E(\bar{y}_\mathbf{S} \mid D = d) = \frac{\Sigma_s y_k}{\nu(s)}$$

Thus the mean of the distinct units in **S**,

$$t^*(D) = \bar{y}_{\mathbf{S}} = \frac{\Sigma_{\mathbf{S}} y_k}{\nu(\mathbf{S})}$$

is, according to Theorem 2.4, an unbiased estimator of $\bar{y}$ with

$$V(\bar{y}_{\mathbf{S}}) \leqslant V(\bar{y}_S) \tag{3.2}$$

The inequality (3.2) was proved independently by Basu (1958) and Raj and Khamis (1958). Pathak (1962) gave the following expression for the variance of $\bar{y}_{\mathbf{S}}$:

$$V(\bar{y}_{\mathbf{S}}) = \frac{E\{1/\nu(S) - 1/N\} N\sigma^2}{(N-1)} = \frac{\sigma^2 \sum_{j=1}^{N-1} (j/N)^{n-1}}{(N-1)}$$

Since $P\{\nu(S) \leqslant n\} = 1$, we have the following inequalities:

$$\frac{(1/n - 1/N)N\sigma^2}{(N-1)} \leqslant \frac{E\{1/\nu(S) - 1/N\}N\sigma^2}{(N-1)} \leqslant \frac{\sigma^2}{n} \tag{3.3}$$

where the smallest term is the variance of the sample mean $\bar{y}_S = \Sigma_S y_k / n$ under the FES($n$) design *srs*. Thus, when comparing *srs* and *srsr*, with $n$ draws in both cases, we see that the strategy (*srs*, $\bar{y}_S$) is better than (*srsr*, $\bar{y}_{\mathbf{S}}$), which is better than (*srsr*, $\bar{y}_S$), as measured by the variance. The inequalities (3.3) are thoroughly discussed by Lanke (1972, 1975). Table 2.1, taken from Lanke (1975), gives the impression that the variance of the strategy (*srsr*, $\bar{y}_{\mathbf{S}}$) is slightly greater than midway between those of the strategies (*srs*, $\bar{y}_S$) and (*srsr*, $\bar{y}_S$).

Note that Rao–Blackwellization in this example did not result in the Horvitz–Thompson estimator, which is also independent of order and multiplicity. Under *srsr*, the inclusion probabilities are $\alpha_k = 1 - (1 - 1/N)^n$

**Table 2.1. Comparison of the Variances $V_1$, $V_2$ and $V_3$ of the Strategies (*srs*, $\bar{y}_S$), (*srsr*, $\bar{y}_{\mathbf{S}}$), and (*srsr*, $\bar{y}_S$), respectively. (From Lanke, 1975.)**

| | $n/N = 0.1$ | | $n/N = 0.2$ | | $n/N = 0.5$ | |
|---|---|---|---|---|---|---|
| $N$ | $V_1/V_3$ | $V_2/V_3$ | $V_1/V_3$ | $V_2/V_3$ | $V_1/V_3$ | $V_2/V_3$ |
| 20 | 0.9474 | 1.0000 | 0.8421 | 0.9500 | 0.5263 | 0.8092 |
| 100 | 0.9091 | 0.9604 | 0.8081 | 0.9123 | 0.5051 | 0.7781 |
| 500 | 0.9018 | 0.9527 | 0.8016 | 0.9051 | 0.5010 | 0.7722 |
| $\infty$ | 0.9000 | 0.9508 | 0.8000 | 0.9033 | 0.5000 | 0.7707 |

$(k = 1, \ldots, N)$. The Horvitz–Thompson estimator of $\bar{y}$ introduced by formula (6.10) of Section 1.6 takes the form

$$t_{HT} = \Sigma_S \frac{y_k}{N\alpha_k} = \frac{\Sigma_S y_k}{N\{1 - (1 - 1/N)^n\}} = \frac{\Sigma_S y_k}{E\{\nu(S)\}} \quad ■$$

**Example 2.** A sequence of $n$ units is drawn by the design *ppsr* as defined in Section 1.4. Formula (6.8) of Section 1.6 defines the unbiased estimator of $\bar{y}$ due to Hansen and Hurwitz (1943) $t_{HH}$, which depends on multiplicity, but not on order, since each unit enters in the formula as many times as it appears in the sequence **S**. Rao–Blackwellization of $t_{HH}$, considered by Pathak (1962), yields a rather complicated estimator which does not admit a simple variance estimator, as does $t_{HH}$. Moreover, the gain in efficiency is considered to be small, unless the sampling fraction is large. Thus, the resulting estimator is less useful in practice than the original Hansen–Hurwitz estimator.

As an illustration we quote an example from Murthy (1967). Suppose that by the design *ppsr* with $n = 3$ we obtain $\mathbf{s} = (1, 2, 1)$. Then $\mathbf{d} = ((1, y_1), (2, y_2), (1, y_1))$ and $d = \{(1, y_1), (2, y_2)\}$. The Hansen–Hurwitz estimator takes the form

$$t_{HH} = \frac{y_1/p_1 + y_2/p_2 + y_1/p_1}{3N} = \frac{2y_1/p_1 + y_2/p_2}{3N}$$

Rao–Blackwellization gives the improved estimator

$$t^* = E\{t_{HH} \mid D = d\} = \sum_{\mathbf{s} \in A_s} \frac{p(\mathbf{s})}{p(s)} t_{HH}$$

$$= \frac{y_1/p_1 + y_2/p_2 + (y_1 + y_2)/(p_1 + p_2)}{3N}$$

where $A_s = \{\mathbf{s} : r(\mathbf{s}) = s\}$.

Note that $t^*$ is not identical to the Horwitz–Thompson estimator, which for $\mathbf{s} = (1, 2, 1)$ takes the form

$$t_{HT} = \Sigma_S \frac{y_k}{N\alpha_k} = \frac{y_1/\{1 - (1 - p_1)^3\} + y_2/\{1 - (1 - p_2)^3\}}{N}$$

since the inclusion probabilities under *ppsr* are $\alpha_k = 1 - (1 - p_k)^n$ $(k = 1, \ldots, N)$. ■

**Example 3.** A sequence of $n$ units is drawn without replacement by the successive sampling design *ppsux* defined in Section 1.4.

In this situation, Raj (1956) suggested the unbiased order-dependent estimator $t_{DR}$ defined by (6.12) in Section 1.6. This estimator has a very simple, and always nonnegative, variance estimator.

Rao–Blackwellization of $t_{DR}$ was considered by Murthy (1957). The resulting unordered estimator can be expressed as

$$t^* = t_{MU} = \frac{\Sigma_S y_k p(S \mid k)}{p(S)} \tag{3.4}$$

where $p(s \mid k)$ denotes the conditional probability of getting the unordered sample $s$, given that unit $k$ was selected in the first draw.

We quote again an example from Murthy (1967). Suppose that a sample is drawn by the design *ppsux* with $n = 2$, resulting in $\mathbf{s} = (1, 2)$, and hence $s = \{1, 2\}$. Then $t_1 = y_1/p_1$ and $t_2 = y_1 + (1 - p_1)y_2/p_2$. The Raj estimator takes the value

$$t_{DR} = \frac{(1 + p_1)y_1/p_1 + (1 - p_1)y_2/p_2}{2N}$$

Rao–Blackwellization gives

$$t^* = t_{MU} = E(t_{DR} \mid D = d) = \sum_{\mathbf{s}' \in A_s} \frac{p(\mathbf{s}')}{p(s)} t_{DR}$$

$$= \frac{(1 - p_2)y_1/p_1 + (1 - p_1)y_2/p_2}{2 - p_1 - p_2}$$

Note that the unordered Raj estimator $t_{MU}$ is not identical to the Horvitz–Thompson estimator, which, for $\mathbf{s} = (1, 2)$, takes the value

$$t_{HT} = \frac{y_1/p_1 \left\{1 + \sum\limits_{k \neq 1}^{N} p_k/(1 - p_k)\right\} + y_2/p_2 \left\{1 + \sum\limits_{k \neq 2}^{N} p_k/(1 - p_k)\right\}}{N}$$

since, for the sampling design under consideration, and for $n = 2$,

$$\alpha_l = p_l \left\{1 + \sum_{k \neq l}^{N} p_k/(1 - p_k)\right\} \qquad (l = 1, 2). \qquad \blacksquare$$

Since, by Theorem 2.2, $D$ is a sufficient statistic, we can restrict our search for good estimators to the class of estimators that depend on $\mathbf{D}$ only through $D = u_0(\mathbf{D})$. But this class contains an infinity of estimators. We still need a device for selecting one unique estimator that would be preferred to all others. Unfortunately, due to the "incompleteness" of the statistic $D$, no such uniquely best estimator exists.

**Definition.** A statistic $Z = u(\mathbf{D})$ is said to be *complete* if and only if, for any function $g(Z)$ with $E\{g(Z)\} = 0$ for all $\mathbf{y} \in R_N$, we have $P\{g(Z) = 0; \mathbf{y}\} = 1$ for all $\mathbf{y} \in R_N$. $\blacksquare$

**Theorem 2.5.** *The statistic $D = u_0(\mathbf{D})$ is not complete.*

**Proof.** We have to find at least one function $g(D)$ such that $E\{g(D)\} = 0$ for all $\mathbf{y} \in R_N$, and $P\{g(D) \neq 0; \mathbf{y}\} > 0$ for at least one $\mathbf{y} \in R_N$.

Consider some fixed $k_0 \in \{1, \ldots, N\}$ such that $0 < \alpha_{k_0} < 1$, and let $g_c(D)$ be defined, for some real constant $c \neq 0$, by

$$g_c(D) = \begin{cases} c/\alpha_{k_0} & \text{if } S \in C_{k_0} \\ -c/(1-\alpha_{k_0}) & \text{if } S \notin C_{k_0} \end{cases}$$

where $C_{k_0} = \{s : k_0 \in s\}$. Then

$$E\{g_c(D)\} = \alpha_{k_0} \frac{c}{\alpha_{k_0}} + (1 - \alpha_{k_0}) \frac{-c}{1 - \alpha_{k_0}} = 0$$

for all $\mathbf{y} \in R_N$. But as soon as $c \neq 0$, we see that $P(g_c(D) \neq 0; \mathbf{y}) = 1$ for all $\mathbf{y} \in R_N$. Thus $D$ is not a complete statistic. □

If $t(D)$ is an unbiased estimator of $\bar{y}$, then $t(D) + g_c(D)$ is also an unbiased estimator of $\bar{y}$. Thus there exist infinitely many unbiased estimators of $\bar{y}$ that are functions of the sufficient statistic $D$.

Note also that in each of Examples 1–3 we constructed an improved estimator $t^*(D)$, not identical to the Horvitz–Thompson estimator $t_{HT} = t_{HT}(D)$. But both $t^*(D)$ and $t_{HT}(D)$ are unbiased, and hence any linear combination $w\, t^*(D) + (1 - w)t_{HT}(D)$ would also be an unbiased estimator of $\bar{y}$, based on $D$.

**Remark 2.** If $D$ had been complete, then there would have existed one unique unbiased estimator based on $D$, which would have been the "uniformly minimum variance unbiased estimator." ■

Summing up, we have concluded that the search for good estimators of $\bar{y}$, as measured by the mean square error, can be limited to those that disregard order and multiplicity. Yet there is an infinity of such estimators, and the results presented so far do not enable us to find a unique best estimator. In fact, we have not yet established whether a best estimator exists or not. This latter question will be considered in some detail in Sections 3.4 and 3.5.

## 2.4. THE LIKELIHOOD FUNCTION

It is easy to identify the appropriate likelihood function under the Fixed population approach. As we shall see, it has a form that makes it a poor tool for inference: No unique maximum likelihood estimate is obtained. However, we shall also see that if we are willing to accept a somewhat different point of departure considered by Royall (1968), and Hartley and Rao (1968, 1969), then another likelihood function emerges, one that readily yields a maximum likelihood estimate of $\bar{y}$ under certain conditions.

Let $d = \{(k, y_k) : k \in s\}$ be the given data, and let $\Omega_d$ be the set of all $\mathbf{y} \in R_N$ such that $d$ is consistent with $\mathbf{y}$, as defined in Section 2.1. Due to the sufficiency of $D$ established in Section 2.2, it is justified to treat $d$, rather than $\mathbf{d}$, as our data.

**Example 1.** Let $N = 4$. Suppose that we have obtained the sample $s = \{1, 2\}$, and that the data are $d = \{(1, 20), (2, 0)\}$. Then $\Omega_d$ is the set of all vectors $\mathbf{y} = (20, 0, y_3, y_4)$, where $-\infty < y_3 < +\infty$ and $-\infty < y_4 < +\infty$. ■

The definition of the likelihood function in survey sampling is analogous to the usual definition of a likelihood function in traditional statistical inference.

**Definition.** For given data $d$, the *likelihood function*, $L_d(\mathbf{y})$, is a function of the parameter $\mathbf{y}$, which, for any $\mathbf{y} \in R_N$, gives the probability of obtaining $d$, if $\mathbf{y}$ were the true parameter value. ■

Thus the likelihood function is given by formula (1.2) for $p_{\mathbf{y}}(d)$, considered as a function of $\mathbf{y}$, for given $d$:

**Theorem 2.6.** *For any design p, the likelihood function is given by*

$$L_d(\mathbf{y}) = \begin{cases} p(s) & \textit{for all } \mathbf{y} \in \Omega_d \\ 0 & \textit{otherwise} \end{cases} \qquad \square \qquad (4.1)$$

The likelihood function (4.1), which was first considered by Godambe (1966), is "flat," equal to $p(s) > 0$, over $\Omega_d$, and zero outside $\Omega_d$, hence lacking a unique maximum. Consequently, there exists no unique maximum likelihood estimate of $\mathbf{y}$. All the likelihood function tells us is that all possible values of the unobserved components of $\mathbf{y}$ in $\Omega_d$ have the same likelihood. Thus the likelihood function (4.1) is "uninformative". Under a superpopulation model, on the other hand, the appropriate likelihood function may be more informative; see, for example, Lindley (1971b) and Royall (1975), as well as the discussion in Section 5.6.

**Example 2.** Let $d$ be as in Example 1, and suppose that $p(s) = 0.2$. The likelihood function is

$$L_d(\mathbf{y}) = \begin{cases} 0.2 & \text{for all } \mathbf{y} \in \Omega_d \\ 0 & \text{otherwise} \end{cases}$$

and all it tells us is that all possible parameter vectors $\mathbf{y} = (20, 0, y_3, y_4)$ are equally likely, where $-\infty < y_3 < +\infty$ and $-\infty < y_4 < +\infty$. This information does not suffice to produce a unique maximum likelihood estimate of $\bar{y} = (20 + 0 + y_3 + y_4)/4$. ■

**Remark 1.** If the parameter space is extremely restricted, a unique maximum likelihood estimate of $\bar{y}$ may exist. Suppose, in Example 2 above, that the parameter space is the set of all vectors $\mathbf{y} \in R_4$ such that $-\infty < y_k < +\infty$ $(k = 1, \ldots, 4)$ and $y_3 + y_4 = 30$. Then, given the data $d = \{(1, 20), (2, 0)\}$, there is a unique maximum likelihood estimate of $\bar{y}$, namely $\bar{y} = (20 + 0 + 30)/4 = 12.5$. The restriction imposed on the parameter space means that we know exactly the sum of certain components of $\mathbf{y}$. Thus, when the remaining components are observed, the population mean $\bar{y}$ becomes completely known. This is, of course, an unrealistic assumption in most practical situations. ∎

**Remark 2.** Survey sampling is not the only area of statistical theory where a flat likelihood function may occur. Basu (1969) gives the following simple example. Let $x_1, \ldots, x_n$ be $n$ independent observations on a continuous random variable $X$ with the density

$$f_\theta(x) = \begin{cases} 1 & \text{for } \theta - 1/2 < x < \theta + 1/2 \\ 0 & \text{otherwise} \end{cases}$$

where $\theta$ is an unknown parameter, and the parameter space is $R_1$. The likelihood function, for given $\mathbf{x} = (x_1, \ldots, x_n)$, is

$$L_{\mathbf{x}}(\theta) = \begin{cases} 1 & \text{for } x_{\max} - 1/2 < \theta < x_{\min} + 1/2 \\ 0 & \text{otherwise} \end{cases}$$

where $x_{\max}$ is the maximum of the observations $x_1, \ldots, x_n$, and $x_{\min}$ is the minimum. Thus $L_{\mathbf{x}}(\theta)$ is flat over the interval $(x_{\max} - 1/2, x_{\min} + 1/2)$, and zero outside. No unique maximum likelihood estimate of $\theta$ exists (unless $x_{\max} - x_{\min} = 1$). ∎

Some writers argue that statistical inference should be guided by the so called *likelihood principle*; see, for example, Barnard, Jenkins, and Winsten (1962) and Birnbaum (1962). The likelihood principle says, roughly, that if two different outcomes (possibly generated under different experimental circumstances) give likelihood functions that differ at most by a multiplicative factor not dependent on the unknown parameters, then the two outcomes should give rise to the same inference. It is seen from (4.1) that the likelihood principle, when applied to survey sampling, under the Fixed population approach, has the consequences (i) and (ii) below:

(i) *Inference from survey data should be independent of the sampling design*. Let $p_0$ and $p_1$ be two designs such that $p_0(s) > 0$ and $p_1(s) > 0$ for a certain sample $s$. For given data $d = \{(k, y_k) : k \in s\}$, under the design $p_i$ $(i = 0, 1)$, the likelihood function would be, say,

$L_d(\mathbf{y}) = L_{d,p_i}(\mathbf{y})$, where, for $i = 0, 1$,

$$L_{d,p_i}(\mathbf{y}) = \begin{cases} p_i(s) & \text{for } \mathbf{y} \in \Omega_d \\ 0 & \text{otherwise} \end{cases}$$

Hence, $L_{d,p_0}(\mathbf{y}) = \{p_0(s)/p_1(s)\}L_{d,p_1}(\mathbf{y})$ for all $y \in R_N$, that is, the two likelihood functions differ only by a multiplicative factor, $p_0(s)/p_1(s)$, not dependent on the parameter vector $\mathbf{y}$. Thus, according to the likelihood principle, the data $d$ should give the same inference about $\mathbf{y}$ irrespective of the design by which the sample was drawn.

(ii) *The only inference about* $\mathbf{y}$ *sanctioned by the likelihood principle is the trivial one that the components* $y_k$ *for* $k \in s$ *must coincide with their observed values.* It follows from the likelihood principle that all essential information from the data is carried by the likelihood function. And we have just seen that the likelihood function arising from the Fixed population approach to survey sampling does not admit discrimination among the possible values of the unobserved components of $\mathbf{y}$, since all values $\mathbf{y} \in \Omega_d$ have the same likelihood.

The two conclusions (i) and (ii) are rather controversial. Most survey statisticians still think that the choice of estimator should be directed by considerations involving the sampling design, for example, that the estimator should be unbiased or have a small MSE under the given design. Moreover, many estimators in common use, for example the Horvitz–Thompson estimator, depend functionally on the design. (However, if the statistician invokes certain model assumptions, it becomes natural to make the choice of estimator regardless of the design; see Chapters 5 and 6.) Also, it is generally felt by survey statisticians that the survey data in some way have more information to give about the population than the trivial information mentioned in (ii) above. In summary, strict adherence to the likelihood principle is inconsistent with widely accepted opinions of sample survey practitioners, many of whom feel that such derivatives of randomization as design unbiasedness and reduction of design variance are extremely important.

In the rest of this section we consider a slightly different approach to the inference problem, proposed independently by Royall (1968) and Hartley and Rao (1968, 1969), and also considered by C. R. Rao (1971). The salient feature of this approach is that the label part of the data is *deliberately* ignored, once the sample has been drawn. It is decided, in other words, that the only data to be observed are the $y_k$-values of the sampled units, *not* their labels. Our notation for these unlabeled data is $y_s = \{y_k : k \in s\}$. Of course, the statistician, in adopting this behavior, must feel convinced that he has but little information to lose by dropping the labels.

**Remark 3.** The deliberate ignorance of labels might be justified in certain cases, for example, (a) when in fact no labels are attached to the units; and (b) when there is no evidence of a relationship between the labels $k$ and the corresponding $y_k$-values; see the discussion in Hartley and Rao (1968). Disregarding labels in case (b) is a controversial issue that has been challenged by Godambe (1970, 1975). His point is essentially that if the labels are available, they must not be ignored. We know that if the label part $s$ of the data is dropped, then the remaining part $y_s$ is no longer sufficient for $\mathbf{y}$. Hence, some information indeed seems to be lost when labels are ignored. On the other hand, it seems futile to try to incorporate facts (in this case, information about the labels) that have no structured relationship to the $y$-values, and hence no apparent inferential value. We end up, for example, with negative results such as those to be discussed in Section 3.4 stating that no UMV estimator exists in certain classes of label-dependent estimators.

A constructive question (to which as yet no satisfactory answer has been given) is: How should labels be utilized for improving an estimator, when we have no knowledge about the type of relationship between the labels and the $y_k$-values? That is, if $t$ is any estimator, not dependent on labels, can we find an estimator $t^*$ which depends on labels and is at least as good as $t$? (One aspect of this question will be discussed in Remarks 3 and 9 of Section 3.1.) ∎

The likelihood function based on the given data $y_s$ can be utilized for inference about $\bar{y}$, at least under certain designs. In the remainder of this section we assume that the design *srs* is used. Let $s$ be a sample obtained by this design. Suppose that the observed unlabeled data $y_s = \{y_k : k \in s\}$ contains exactly $m (\leqslant n)$ *distinct* values, denoted $x_1, \ldots, x_m$, and let $n_h$ be the number of times that $x_h$ appears in $y_s$; $\sum_{h=1}^{m} n_h = n$. Then the probability of getting the observed data $y_s$ is given by the multivariate hypergeometric distribution

$$P(y_S = y_s; \mathbf{y}) = \frac{\prod_{h=1}^{m} \binom{N_h}{n_h}}{\binom{N}{n}}$$

where $N_1, \ldots, N_m$ are the multiplicities of $x_1, \ldots, x_m$ in the population vector $\mathbf{y}$. Thus the likelihood function, for given data $y_s$, is

$$L_{y_s}(\mathbf{y}) = \begin{cases} \prod_{h=1}^{m} \binom{N_h}{n_h} \Big/ \binom{N}{n} & \text{for all } \mathbf{y} \in \mathscr{M} \\ 0 & \text{otherwise} \end{cases} \tag{4.2}$$

where, for any $\mathbf{y} \in R_N$, $N_1, \ldots, N_m$ denote the multiplicities of the observed $x_1, \ldots, x_m$ in $\mathbf{y}$, and $\mathscr{M} = \{\mathbf{y} \in R_N : N_h \geqslant n_h \ (h = 1, \ldots, m)$ and $\sum_{h=1}^{m} N_h \leqslant N\}$.

**Remark 4.** The likelihood given by (4.2), considered by Royall (1968), is an example of what is sometimes called a marginal likelihood, as defined, for example, in Cox and Hinkley (1974). ■

The likelihood function (4.2) is not flat, at least not in the same way as (4.1). Often (but not always) the set of vectors **y** that maximize (4.2), makes all $N_h$ uniquely determined. If (4.2) is maximized by $N_h = \hat{N}_h$ $(h = 1, \ldots, m)$, then $\Sigma_{h=1}^m \hat{N}_h = N$. For, if $\Sigma_{h=1}^m N_h < N$, then (4.2) can always be increased by increasing the $N_h$.

In the case where the $\hat{N}_h$ that maximize the likelihood are uniquely determined, we obtain a maximum likelihood estimate of $\bar{y}$ as

$$\hat{\bar{y}} = \frac{\Sigma_{h=1}^m \hat{N}_h x_h}{N} \tag{4.3}$$

When $N/n$ is an integer, (4.2) is maximized for $\hat{N}_h = Nn_h/n$, which makes the maximum likelihood estimate $\hat{\bar{y}}$ equal to the sample mean $\bar{y}_s$.

When $N/n$ is not an integer, the $\hat{N}_h$ that maximize (4.2) are found by rounding off the numbers $Nn_h/n$. The resulting estimator (4.3) is not necessarily equal to the sample mean $\bar{y}_s$. Hartley and Rao (1968, 1969) give an algorithm for computing the $\hat{N}_h$ in the case where $N/n$ is not an integer.

**Example 3.** Assume that the design *srs* has produced a sample such that, for two units, $y_k = 10$, and, for one unit, $y_k = 20$. Labels are ignored and we observe only $y_s = \{10, 10, 20\}$. The distinct sample values, $x_1 = 10$ and $x_2 = 20$, have the multiplicities $n_1 = 2$ and $n_2 = 1$. The likelihood function is, assuming $N = 7$,

$$L_{y_s}(\mathbf{y}) = \begin{cases} \binom{N_1}{2}\binom{N_2}{1}\Big/\binom{7}{3} & \text{for all } \mathbf{y} \in R_7 \text{ such that } N_1 \geqslant 2 \\ & N_2 \geqslant 1 \text{ and } N_1 + N_2 \leqslant 7 \\ 0 & \text{otherwise} \end{cases} \tag{4.4}$$

We compute the value of (4.4) for all $N_1$ and $N_2$ such that $N_1 \geqslant 2, N_2 \geqslant 1$ and $N_1 + N_2 = 7$, to see when it attains its maximum. (As mentioned above, we need not worry about the case where $N_1 + N_2 < 7$.) For $N_1 = 2, 3, 4, 5$, and 6, (4.4) takes the value 5/35, 12/35, 18/35, 20/35, and 15/35, respectively. Hence the maximum occurs for $N_1 = 5 = \hat{N}_1$ and $N_2 = 2 = \hat{N}_2$, which gives the maximum likelihood estimate of $\bar{y}$ as

$$\hat{\bar{y}} = \frac{\hat{N}_1 x_1 + \hat{N}_2 x_2}{N} = \frac{5 \times 10 + 2 \times 20}{7} = 12.86$$

This estimate differs from the sample mean,

$$\bar{y}_s = \frac{n_1 x_1 + n_2 x_2}{n} = \frac{2 \times 10 + 20}{3} = 13.33$$

■

**Remark 5.** One objection raised by Kempthorne (1969) against the maximum likelihood estimates $\hat{N}_h$ $(h = 1, \ldots, m)$ is the following. When accepting $\hat{N}_h$ $(h = 1, \ldots, m)$ as reasonable estimates, we also accept, since $\sum_{h=1}^{m} \hat{N}_h = N$, that the population contains no other distinct values than the observed $x_1, \ldots, x_m$. In many situations such an assumption is unrealistic. Often it is known that the actual population contains distinct values other than those represented in the sample, even though one may not know what those other values are. This objection points to the difficulty of specifying the most appropriate parameter space in the survey sampling setup. ■

**Remark 6.** Hartley and Rao (1968, 1969) assume the population values $y_1, \ldots, y_N$ to be measured on a known scale with a finite number of *scale points* $x_1, \ldots, x_M$, $(M \leqslant N)$, known in advance. They assume the population to be described by a parameter vector $\mathbf{N} = (N_1, \ldots, N_M)$, where $N_h$ is the multiplicity of the value $x_h$ in the population. Thus many $N_h$ may be zero. The population mean can be expressed as

$$\bar{y} = \sum_{h=1}^{M} \frac{N_h x_h}{N}$$

They derive maximum likelihood estimates $\hat{\mathbf{N}} = (\hat{N}_1, \ldots, \hat{N}_M)$ and obtain the maximum likelihood estimate of $\bar{y}$ as

$$\hat{\bar{y}} = \sum_{h=1}^{M} \frac{\hat{N}_h x_h}{N}$$

The results of Hartley and Rao coincide with the results given earlier in this section. For the scale-points $x_h$ not represented in the sample, their estimates are $\hat{N}_h = 0$; for those $x_h$ that have nonzero frequency in the sample, their estimate is the set of $\hat{N}_h$ that maximizes (4.2). Certain results on uniformly minimum variance estimation can also be derived from the "scale load approach" of Hartley and Rao (1968); see Remark 3 of Section 3.5, which also comments on the use of this approach for designs other than *srs*. ■

**Remark 7.** Earlier in this section we saw that the special shape of the likelihood function (4.1) resulted in certain anomalies when the likelihood principle was invoked. A related anomaly is observed when we try to apply the concepts of ancillarity and conditional inference, as described, for example, by Basu (1964).

It is sometimes argued that inference is to be made conditional on some ancillary statistic, that is, a statistic with a probability distribution that does not depend on the unknown parameter. In survey sampling $S$ is an ancillary statistic. But the conditional distribution of $D$, given $S = s$, is a degenerate distribution with all probability concentrated in one point, $d = \{(k, y_k)\colon k \in s\}$. Hence no meaningful inference seems possible conditional on $S$. For example, the conditional variance of any estimator is 0, for all $s \in \mathscr{S}$, $V\{t(D) \mid S = s\} = 0$. Another ancillary statistic in survey sampling is the effective sample size $\nu(S)$. The above mentioned anomaly does not appear when inference is made conditional on $\nu(S)$. Such inference is in fact often made in sample survey practice. ■

CHAPTER 3

# Inference under the Fixed Population Model: Criteria for Judging Estimators and Strategies

A great deal of recent research into the foundations of survey sampling has been devoted to the application of various inference criteria. Well-known criteria from traditional statistical inference have been investigated, such as admissibility and minimum variance. Certain new criteria have also been suggested with exclusive reference to the conditions of survey sampling. One example of the latter kind is the concept of hyperadmissibility (Hanurav, 1968).

This type of research was initiated by Godambe (1955), who demonstrated that there exists no linear estimator of $\bar{y}$ with uniformly minimum variance. Much of the ensuing work has been motivated by the wish to state precisely what favorable properties, if any, could be claimed for the estimators and strategies commonly used in established practice. Through these efforts considerable insight has been gained into the structure of the inference problem for survey populations, and it has become evident that survey sampling inference is not just a straightforward application of traditional statistical inference, as was commonly believed earlier. In this chapter we present some important findings relating to admissibility, hyperadmissibility, minimum variance, minimaxity, and invariance.

## 3.1. ADMISSIBILITY OF ESTIMATORS

Admissibility in survey sampling is defined as in traditional statistical inference. The mean square error will serve as a risk function in the discussion below. Admissibility of an unbiased estimator is a weaker claim than uniformly minimum variance (UMV); see Sections 3.4–3.5. The reason why the admissibility criterion has attracted considerable attention in the theory of survey sampling is related to the nonexistence, within some general classes of estimators, of a UMV unbiased estimator, see Section 3.4. The criterion of admissibility, like sufficiency, does not single out one unique estimator. Many traditional estimators in survey sampling have been shown to be admissible.

Admissibility in survey sampling was first considered by Godambe (1960) and Roy and Chakravarti (1960). Important contributions are given in Godambe and Joshi (1965) and in a series of papers by Joshi (1965a, b, 1966, 1968, 1969).

Admissibility of an estimator $t = t(\mathbf{D})$ or $t = t(D)$ is defined in the following way:

**Definition.** An estimator $t_0$, belonging to some class $\mathscr{C}$ of estimators, is said to be *admissible* in $\mathscr{C}$ under a given design $p$, if and only if no estimator in $\mathscr{C}$ is better than $t_0$. (See Remark 1 of Section 2.3 for the definition of "better.") ■

For each inadmissible estimator in the class $\mathscr{C}$ we can find at least one better estimator among the admissible ones in $\mathscr{C}$. Hence we can restrict the search for estimators with small MSE to the set of admissible estimators in $\mathscr{C}$. However, in judging the admissibility property from a practical point of view, we must not lose sight of the fact that an inadmissible estimator which is just "slightly inadmissible" may have so many practical advantages that it is actually preferred over an admissible one. Therefore, we shall have reason, in Chapter 7, to reconsider certain inadmissible estimators.

**Remark 1.** The following is an immediate consequence of Theorem 2.4 in Section 2.3: Any estimator $t$ that depends on order and/or multiplicity of the sampled units is inadmissible, since the Rao–Blackwellization $t^* = E(t \mid D)$ is better than $t$ (provided that both $t$ and $t^*$ belong to the class $\mathscr{C}$ under consideration). ■

The admissibility of the generalized difference estimator is established in the following Theorem 3.1, from which we obtain, as corollaries, certain admissibility results for the Horvitz–Thompson estimator.

First, we state some preliminaries. The notation for the classes of estimators to be considered is summarized in the following tableau (see Chapter 1 for the exact definitions).

| | Biased as well as Unbiased | Unbiased |
|---|---|---|
| All estimators | $\mathscr{A}$ | $\mathscr{A}_u$ |
| All linear (homogeneous and nonhomogeneous) estimators | $\mathscr{L}$ | $\mathscr{L}_u$ |
| All linear homogeneous estimators | $\mathscr{L}_0$ | $\mathscr{L}_{0u}$ |

**Remark 2.** It is tacitly understood that the classes $\mathscr{A}_u$, $\mathscr{L}_u$, and $\mathscr{L}_{0u}$ are defined with reference to a *given* design, since an estimator that is unbiased under one design is not necessarily so under another design. A more precise notation would have been $\mathscr{A}_u(p)$, $\mathscr{L}_u(p)$, and $\mathscr{L}_{0u}(p)$. ■

The generalized difference estimator $t_{GD}$ was defined, for a predetermined vector $\mathbf{e} = (e_1, \ldots, e_N) \in R_N$, by formula (6.14) of Section 1.6. It is an unbiased estimator of $\bar{y}$, under any design such that $\alpha_k > 0$ $(k = 1, \ldots, N)$. Theorem 3.1 below shows that $t_{GD}$ is admissible in $\mathscr{A}_u$. The proof is based on the following lemma, which is a generalization of a lemma given by Godambe and Joshi (1965). We make use of a special partition of the parameter space $R_N$. For a given vector $\mathbf{e} \in R_N$, define $\Omega_m$ $(m = 0, 1, \ldots, N)$ as

$$\Omega_m = \{\mathbf{y} \in R_N : y_k \neq e_k \text{ for exactly } m \text{ components of } \mathbf{y}\}$$

Clearly, $\{\Omega_0, \Omega_1, \ldots, \Omega_N\}$ is a partition of $R_N$, and $\Omega_0 = \{\mathbf{e}\}$.

**Lemma 3.1.** *If, for any given design with $\alpha_k > 0$ $(k = 1, \ldots, N)$, $t$ is an estimator in $\mathscr{A}_u$ satisfying conditions (a) and (b):*

*(a)* $V(t) \leqslant V(t_{GD})$ *for all* $\mathbf{y} \in R_N$

*(b)* $t(d) = t_{GD}(d)$ *for all* $s \in \mathscr{S}$ *with* $p(s) > 0$, *when* $\mathbf{y} \in \Omega_m$

*then* $t(d) = t_{GD}(d)$ *for all* $s \in \mathscr{S}$ *with* $p(s) > 0$, *when* $\mathbf{y} \in \Omega_{m+1}$.

**Proof.** Let $\mathbf{y}_0 = (y_{01}, \ldots, y_{0N})$ be an arbitrary vector in $\Omega_{m+1}$. Let $\mathscr{S}_j$ $(j = 0, 1, \ldots, m+1)$ be defined as

$$\mathscr{S}_j = \{s \in \mathscr{S} : y_{0k} \neq e_k \text{ for exactly } j \text{ labels } k \in s\}$$

where $\mathbf{e} = (e_1, \ldots, e_N)$ is the constant vector on which $t_{GD} = \Sigma_s (y_k - e_k)/N\alpha_k + \bar{e}$ depends. Any outcome, $d = \{(k, y_k): k \in s\}$, that can occur when $s \in \cup_{j=0}^{m} \mathscr{S}_j$ and $\mathbf{y} = \mathbf{y}_0$ can also occur when $\mathbf{y} \in \Omega_m$, because such an outcome contains at most $m$ $y_k$-values not equal to $e_k$. Hence from condition (b) of the lemma, for $\mathbf{y} = \mathbf{y}_0$,

$$t(d) = t_{GD}(d) \tag{1.1}$$

for all $s \in \cup_{j=0}^{m} \mathscr{S}_j$ with $p(s) > 0$. Note that $t_{GD}(d)$ is constant for all $s \in \mathscr{S}_{m+1}$, when $\mathbf{y} = \mathbf{y}_0$, because then every $s \in \mathscr{S}_{m+1}$ contains those $m+1$ labels, $k$, for which $y_k - e_k \neq 0$. Hence, for $\mathbf{y} = \mathbf{y}_0$,

$$t_{GD}(d) = c \tag{1.2}$$

for all $s \in \mathscr{S}_{m+1}$ with $p(s) > 0$.

Since $t$ is unbiased for $\bar{y}$, we have from (1.1) that, for $\mathbf{y} = \mathbf{y}_0$,

$$E(t - t_{GD}) = \sum_{s \in \mathscr{S}_{m+1}} p(s)(t - t_{GD}) = 0 \tag{1.3}$$

From (1.1) and condition (a) of the lemma, for $\mathbf{y} = \mathbf{y}_0$,

$$V(t) - V(t_{GD}) = \sum_{s \in \mathscr{S}_{m+1}} p(s)(t^2 - t_{GD}^2) \leqslant 0 \tag{1.4}$$

Now, from (1.1)–(1.4) it follows that, for $\mathbf{y} = \mathbf{y}_0$,

$$\sum_{s \in \mathscr{S}} p(s)(t - t_{GD})^2$$

$$= \sum_{s \in \mathscr{S}_{m+1}} p(s)(t^2 - t_{GD}^2) + 2 \sum_{s \in \mathscr{S}_{m+1}} p(s) t_{GD}(t_{GD} - t)$$

$$\leqslant 2 \sum_{s \in \mathscr{S}_{m+1}} p(s) t_{GD}(t_{GD} - t) = 2c \sum_{s \in \mathscr{S}_{m+1}} p(s)(t_{GD} - t) = 0$$

This means that $t(d) = t_{GD}(d)$ for all $s \in \mathscr{S}$ with $p(s) > 0$, when $\mathbf{y} = \mathbf{y}_0$. Since $\mathbf{y}_0$ was arbitrarily chosen from $\Omega_{m+1}$, the lemma is proved. □

Now we can easily prove the admissibility of $t_{GD}$.

**Theorem 3.1.** *For any given design with $\alpha_k > 0$ $(k = 1, \ldots, N)$, and for any given vector $\mathbf{e} \in R_N$, the generalized difference estimator $t_{GD} = \Sigma_s(y_k - e_k)/N\alpha_k + \bar{e}$, is admissible in the class $\mathscr{A}_u$ of all unbiased estimators of $\bar{y}$.*

**Proof.** Suppose that $t$ is an unbiased estimator of $\bar{y}$, such that

$$V(t) \leqslant V(t_{GD}), \text{ all } \mathbf{y} \in R_N \tag{1.5}$$

When $\mathbf{y} \in \Omega_0$, that is, when $\mathbf{y} = \mathbf{e}$, we have that $t_{GD}(d) = \bar{e}$ for all $s \in \mathscr{S}$, which implies

$$V(t_{GD}) = 0 \text{ for } \mathbf{y} \in \Omega_0 \tag{1.6}$$

Now, (1.5) and (1.6) imply that $V(t) = 0$ for $\mathbf{y} \in \Omega_0$. Furthermore, $E(t) = \bar{e}$ for $\mathbf{y} \in \Omega_0$. Hence, $t(d) = \bar{e} = t_{GD}(d)$ for all $s \in \mathscr{S}$ with $p(s) > 0$, when $\mathbf{y} \in \Omega_0$. By repeated use of Lemma 3.1, we get $t(d) = t_{GD}(d)$ for all $s \in \mathscr{S}$ with $p(s) > 0$, when $\mathbf{y} \in \Omega_1, \Omega_2, \ldots, \Omega_N$, that is, $t(d) = t_{GD}(d)$ for all $s \in \mathscr{S}$ with $p(s) > 0$, and for all $\mathbf{y} \in R_N$. Thus, for any estimator $t \in \mathscr{A}_u$ such that $V(t) \leqslant V(t_{GD})$ for all $\mathbf{y} \in R_N$, it is impossible to have $V(t) < V(t_{GD})$ for any $\mathbf{y} \in R_N$. Therefore, $t_{GD}$ is admissible in $\mathscr{A}_u$. □

By considering special values of the constant vector $\mathbf{e}$, we obtain various corollaries of Theorem 3.1: When $\mathbf{e} = \mathbf{0}$, $t_{GD}$ is equal to $t_{HT}$, whence the following statement due to Godambe and Joshi (1965):

**Corollary 3.1.** *For any given design with $\alpha_k > 0$ $(k = 1, \ldots, N)$, the Horvitz–Thompson estimator $t_{HT} = \Sigma_s y_k / N\alpha_k$ is admissible in the class $\mathscr{A}_u$ of all unbiased estimators of $\bar{y}$.* □

By considering $\mathbf{e} = c\mathbf{x}$, where $\mathbf{x}$ is a known vector of auxiliary variable values, and $c$ is a predetermined constant, we obtain a second conclusion:

**Corollary 3.2.** *For any given design with $\alpha_k > 0$ $(k = 1, \ldots, N)$, and for any real constant $c$, the particular form of the generalized difference estimator $t_{GD}$ expressed by (6.15) of Section 1.6 is admissible in the class $\mathcal{A}_u$ of all unbiased estimators of $\bar{y}$.* □

**Example 1.** Under the design *srs*, it follows immediately from Corollaries 3.1 and 3.2 that both the sample mean $\bar{y}_S$ and the difference estimator $t_D = \bar{y}_S + c(\bar{x} - \bar{x}_S)$ are admissible in $\mathcal{A}_u$. ■

**Remark 3.** Since, under the design *srs*, $\bar{y}_S$ is admissible in $\mathcal{A}_u$, there exists no unbiased estimator of $\bar{y}$ that is better than $\bar{y}_S$. Moreover, the value of $\bar{y}_S$ can be computed without knowing the labels of the sampled units. Hence the following conclusion: Knowledge of the labels of the sampled units does not, under *srs*, enable us to construct an unbiased estimator of $\bar{y}$ that is better than $\bar{y}_S$.

Moreover, we see in Section 3.5, Corollary 3.8, that $\bar{y}_S$ is, under the design *srs*, the only unbiased estimator that does not depend on the labels.

The estimator $t_D$ is also admissible in $\mathcal{A}_u$ under the design *srs*, but $t_D$ is, in contrast to $\bar{y}_S$, label-dependent. ■

**Remark 4.** From the definition of admissibility we conclude: If $t$ is an estimator, and $\mathcal{C}_1$ and $\mathcal{C}_2$ are classes of estimators such that $t \in \mathcal{C}_1, t \in \mathcal{C}_2$ and $\mathcal{C}_1 \subseteq \mathcal{C}_2$, and

(i) if $t$ is admissible in $\mathcal{C}_2$, then $t$ is also admissible in $\mathcal{C}_1$; that is, admissibility extends to the smaller class;

(ii) if $t$ is inadmissible in $\mathcal{C}_1$, then $t$ is also inadmissible in $\mathcal{C}_2$; that is, inadmissibility extends to the larger class.

From Theorem 3.1 and Corollary 3.1 we obtain the following results, using rule (i) above and the fact that $\mathcal{L}_{0u} \subset \mathcal{L}_u \subset \mathcal{A}_u$:

(a) For any given design with $\alpha_k > 0$ $(k = 1, \ldots, N)$, $t_{GD}$ is admissible in the class $\mathcal{L}_u$ of all unbiased linear estimators of $\bar{y}$.

(b) For any given design with $\alpha_k > 0$ $(k = 1, \ldots, N)$, $t_{HT}$ is admissible in the class $\mathcal{L}_u$ of all unbiased linear estimators of $\bar{y}$.

(c) For any given design with $\alpha_k > 0$ $(k = 1, \ldots, N)$, $t_{HT}$ is admissible in the class $\mathcal{L}_{0u}$ of all unbiased linear homogeneous estimators of $\bar{y}$.

Result (c) was proved independently by Godambe (1960) and Roy and Chakravarti (1960). ■

**Remark 5.** In the proof of Theorem 3.1, it is tacitly understood that the parameter space equals $R_N$. The same proof is, however, valid for more restricted parameter spaces. The crucial point in the proof is that **e** belongs to the parameter space. Thus, the proof of Theorem 3.1 establishes the admissibility of $t_{GD}$ for any parameter space that is a subset of $R_N$ and contains **e**, for example, the space consisting of all $\mathbf{y} \in R_N$ such that $b_k \leqslant y_k \leqslant c_k$ $(k = 1, \ldots, N)$, where $b_k \leqslant e_k \leqslant c_k$.

For Corollary 3.1 to hold, it is essential that **0** belongs to the parameter space. Thus, the admissibility of $t_{HT}$ follows, for example, when the parameter space is:

(a) the set of all $\mathbf{y} \in R_N$ such that $b_k \leqslant y_k \leqslant c_k$ $(k = 1, \ldots, N)$ where $b_k \leqslant 0 \leqslant c_k$;
(b) the set of all $\mathbf{y} \in R_N$ such that $y_k$ is 0 or 1.

Case (b) covers a type of population often encountered in sampling practice, where 0 and 1 denote the absence and the presence, respectively, of a certain property. ■

So far we have only considered classes of *unbiased* estimators. We have seen that $t_{GD}$ and $t_{HT}$ are admissible in $\mathscr{A}_u$, and in $\mathscr{L}_u$. Moreover we found that $t_{HT}$ is admissible in $\mathscr{L}_{0u}$. In Table 3.1 we present, without proofs, some results

**Table 3.1. Some further Results on Admissibility**

| Result | Design | Class of Estimators | Details of Result |
|---|---|---|---|
| A | Any given design | $\mathscr{A}$ | The sample mean of distinct units $\bar{y}_S = \Sigma_S y_k / \nu(S)$ is admissible (Joshi, 1965a). |
| B | Any given design | $\mathscr{A}$ | The ratio estimator $t_R = \bar{x}\bar{y}_S/\bar{x}_S$ is admissible (Joshi, 1966). |
| C | Any given FES($n$) design | $\mathscr{A}$ | All estimators of type $t = \Sigma_S b_k y_k$, where (1) $b_k \geqslant 1/N$ $(k = 1, \ldots, N)$ (2) $b_k = 1/N \Rightarrow \alpha_k = 1$ (3) $\Sigma_1^N 1/b_k = nN$ are admissible (Joshi, 1965b, 1966). |
| D | Any given non-FES design | $\mathscr{L}_0$ | The Horvitz–Thompson estimator $t_{HT} = \Sigma_S y_k / N\alpha_k$ is inadmissible (Godambe and Joshi, 1965). |

due to Joshi on admissibility when the assumption of unbiasedness is dropped. Then we utilize Table 3.1 for some additional conclusions concerning admissibility.

**Remark 6.** By Remark 4 above, case (i), Results A, B, and C in Table 3.1 can be extended to the class $\mathscr{L}$ as well as to the class $\mathscr{L}_0$. ■

**Remark 7.** From Result C in Table 3.1 it follows that the Horvitz–Thompson estimator is admissible in the class $\mathscr{A}$ of all estimators, for any given FES design. We can express $t_{HT}$ as $t_{HT} = \Sigma_S b_k y_k$ with $b_k = 1/N\alpha_k$ $(k = 1, \ldots, N)$, and it is easily seen that the conditions (1)–(3) of Result C are satisfied. ■

**Remark 8.** Using case (ii) of Remark 4 above, we see that Result D in Table 3.1 implies that $t_{HT}$ is inadmissible in the class $\mathscr{A}$ of all estimators, for any given non-FES design. ■

**Example 2.** The design *srsr* is non-FES. We know from Section 2.4 that the mean of $y$ in the $n$ draws $\bar{y}_S$ is inadmissible in $\mathscr{A}_u$, and in $\mathscr{A}$, because a better estimator is the mean of the distinct units $\bar{y}_S$ obtained from $\bar{y}_S$ by Rao–Blackwellization. We also noted, in Example 1 of Section 2.3, that the Horvitz–Thompson estimator under *srsr* differs from $\bar{y}_S$, but like $\bar{y}_S$ it is independent of order and multiplicity. Then a natural question is: Which of the two estimators, $\bar{y}_S$ or $t_{HT}$, would be preferred under *srsr*? The criterion of admissibility does not give much guidance. From the results of the present section we can conclude:

(a) $\bar{y}_S$ and $t_{HT}$ are both admissible in $\mathscr{A}_u$, under *srsr*;
(b) $\bar{y}_S$ is admissible in $\mathscr{A}$ but $t_{HT}$ is inadmissible in $\mathscr{A}$, under *srsr*.

Statement (a) follows from Corollary 3.1 and from Result A in Table 3.1, since $\mathscr{A}_u \subset \mathscr{A}$. Statement (b) follows from Result A and from Remark 8 above. ■

**Remark 9.** Result A of Table 3.1 tells us that, under any given design, no estimator in $\mathscr{A}$ is better than $\bar{y}_S$. Since the value of $\bar{y}_S$ can be computed without knowing the labels of the sampled units, we conclude: For any given design, knowledge of the labels of the sampled units does not enable us to construct an estimator in $\mathscr{A}$ that is better than $\bar{y}_S$. (Compare Remark 3 of Section 2.4.)

But it is not known whether $\bar{y}_S$ is the only estimator in $\mathscr{A}$ that is admissible and does not depend on the labels.

Note that if consideration is limited to the subclass $\mathscr{A}_u$ of all unbiased estimators of $\bar{y}$, then the results of this section tell us that, under *any* given design, it is impossible to utilize the labels toward construction of an estimator in $\mathscr{A}_u$ that is better than $\bar{y}_S$. Since $\bar{y}_S$ is admissible in $\mathscr{A}$, and since $\mathscr{A}_u$ is a subclass of $\mathscr{A}$, no estimator in $\mathscr{A}_u$ can be better than $\bar{y}_S$.

Also note that even though no estimator in $\mathscr{A}$ is better than $\bar{y}_S$, there may exist many estimators in $\mathscr{A}$ that have smaller MSE than $\bar{y}_S$ when $\mathbf{y}$ is restricted to certain subspaces of the parameter space $R_N$; see Example 1 of Section 1.5. ■

## 3.2. ADMISSIBILITY OF STRATEGIES

The problem considered in the preceding section was: Can we, under a *given* design $p$, find an estimator that is better than a particular estimator $t$? If the answer was "no," then $t$ was said to be an admissible estimator under the given design $p$.

Now let us consider a natural extension of the foregoing problem, namely: Can we, by changing both design and estimator, find a strategy that is better than a particular strategy $(p, t)$? This question is made more precise through the following definitions.

The concepts of unbiasedness, variance and MSE (that is, in the $p$-sense), with reference to a strategy $(p, t)$, were defined in Section 1.6. A strategy $(p_1, t_1)$ is said to be *at least as good* as a competing strategy $(p_2, t_2)$ if $\text{MSE}(p_1, t_1) \leqslant \text{MSE}(p_2, t_2)$ for all $\mathbf{y} \in R_N$. If, in addition, strict inequality holds for at least one $\mathbf{y} \in R_N$, then the strategy $(p_1, t_1)$ is said to be *better* than $(p_2, t_2)$.

**Definition.** A strategy $(p, t)$ belonging to some class $\mathscr{H}$ of strategies is said to be *admissible* in $\mathscr{H}$, if and only if no strategy in $\mathscr{H}$ is better than $(p, t)$. ■

We shall consider two classes of strategies: the class $\mathscr{H}_u(n)$ of all unbiased strategies with expected effective sample size fixed at $n$, that is, $\Sigma_{\mathscr{S}} p(s)\nu(s) = n$; and the class $\mathscr{H}(n)$ of all (biased and unbiased) strategies with expected effective sample size equal to $n$.

Admissibility of strategies was first considered by Joshi (1966), who called it "uniform admissibility." Various results on admissible strategies are given by Godambe (1969a), Ericson (1970), Scott (1975a), and Sekkappan and Thompson (1975).

Ramakrishnan (1975) proved that any "Horvitz–Thompson strategy" $(p, t_{HT})$ with expected size $n$ is admissible in $\mathscr{H}_u(n)$. We shall give a straightforward generalization of this result by showing, in Theorem 3.2, the admissibility in $\mathscr{H}_u(n)$ of any "generalized difference strategy" $(p, t_{GD})$ where $p$ has expected effective size $n$ but is otherwise arbitrary, and the vector $\mathbf{e} = (e_1, \ldots, e_N)$ that specifies $t_{GD} = \Sigma_S(y_k - e_k)/N\alpha_k + \bar{e}$ is arbitrary.

The proof of Theorem 3.2 requires the following lemma, which involves two fixed designs, $p_0$ and $p_1$, whose inclusion probabilities are denoted $\alpha_k(0)$ and $\alpha_{kl}(0)$ in the case of the former design, and $\alpha_k(1)$ and $\alpha_{kl}(1)$ in the case of the latter.

**Lemma 3.2.** *If $(p_0, t_{GD})$ and $(p_1, t_1)$ are strategies in $\mathscr{H}_u(n)$ satisfying*

$$V(p_1, t_1) \leqslant V(p_0, t_{GD}) \quad \textit{for all} \quad \mathbf{y} \in R_N \tag{2.1}$$

*then $\alpha_k(0) = \alpha_k(1)$ and $\alpha_{kl}(0) = \alpha_{kl}(1)$ for $k \neq l = 1, \ldots, N$.*

**Proof.** Let $(p_0, t_{GD})$ and $(p_1, t_1)$ be two arbitrary strategies in $\mathscr{H}_u(n)$, satisfying condition (2.1). This implies that, for all $\mathbf{y} \in R_N$,

$$E_{p_1}\{(t_1 - \bar{e})^2\} \leqslant E_{p_0}\{(t_{GD} - \bar{e})^2\} \tag{2.2}$$

where, for any $p$ and $t$, the operator $E_p$ is used as follows: $E_p(t) = \Sigma_{\mathscr{S}} p(s)t$.

First, consider $\mathbf{y} = \mathbf{e}$. Then $t_{GD} - \bar{e} = 0$ for all $s \in \mathscr{S}$. From (2.2) it follows that $E_{p_1}\{(t_1 - \bar{e})^2\} = 0$. Hence for $\mathbf{y} = \mathbf{e}$,

$$t_1 - \bar{e} = 0 \quad \text{for all } s \quad \text{with} \quad p_1(s) > 0 \tag{2.3}$$

Secondly, consider a vector $\mathbf{y} = \mathbf{y}' = (y_1', \ldots, y_N')$ such that $\mathbf{y}' - \mathbf{e}$ has exactly one nonzero component, say $y_k' - e_k \neq 0$. It follows from (2.3) that $t_1 - \bar{e} = 0$ for all $s$ such that $p_1(s) > 0$ and such that the $y$-value agrees with the $e$-value for every label in $s$. That is, when $\mathbf{y} = \mathbf{y}'$, we have $t_1 - \bar{e} = 0$ for all $s \notin C_k = \{s : k \in s\}$ such that $p_1(s) > 0$. Moreover, $t_1$ is unbiased for $\bar{y}$, so that, for $\mathbf{y} = \mathbf{y}'$,

$$E_{p_1}(t_1 - \bar{e}) = E^k_{p_1}(t_1 - \bar{e}) = \frac{y_k' - e_k}{N} \tag{2.4}$$

where, for arbitrary $k, p$, and $t$, the new operator $E^k_p$ is used as follows:

$$E^k_p(t) = \Sigma_{C_k} p(s)t$$

Using the Cauchy–Schwartz inequality and noting that $\alpha_k(j) = \Sigma_{C_k} p_j(s)$ for $j = 0, 1$, we have

$$\{E^k_{p_1}(t_1 - \bar{e})\}^2 \leqslant \alpha_k(1) E^k_{p_1}\{(t_1 - \bar{e})^2\} = \alpha_k(1) E_{p_1}\{(t_1 - \bar{e})^2\} \tag{2.5}$$

Since for $\mathbf{y} = \mathbf{y}'$,

$$E_{p_0}\{(t_{GD} - \bar{e})^2\} = \frac{(y_k' - e_k)^2}{N^2 \alpha_k(0)} \tag{2.6}$$

we obtain from (2.2), (2.4), (2.5), and (2.6)

$$\frac{(y_k' - e_k)^2}{N^2 \alpha_k(1)} \leqslant \frac{(y_k' - e_k)^2}{N^2 \alpha_k(0)}$$

which implies

$$\alpha_k(0) \leqslant \alpha_k(1) \tag{2.7}$$

But (2.7) holds for any $k = 1, \ldots, N$. Since $\Sigma_1^N \alpha_k(0) = \Sigma_1^N \alpha_k(1) = n$, we conclude that

$$\alpha_k(0) = \alpha_k(1) = \alpha_k, \quad \text{say } (k = 1, \ldots, N)$$

It now follows, for $\mathbf{y} = \mathbf{y}'$, that $V(p_1, t_1) = V(p_0, t_{GD}) = (y_k' - e_k)^2/N^2\alpha_k$ and that equality holds in (2.5), so that

$$t_1 - \bar{e} = \frac{y_k' - e_k}{N\alpha_k} \quad \text{for all } s \in C_k \text{ with } p_1(s) > 0 \tag{2.8}$$

Thirdly, consider a vector $\mathbf{y} = \mathbf{y}'' = (y_1'', \ldots, y_N'')$ such that $\mathbf{y}'' - \mathbf{e}$ has exactly two nonzero components, say $y_k'' - e_k \neq 0$ and $y_l'' - e_l \neq 0$. When $\mathbf{y} = \mathbf{y}''$, it follows from (2.3) and (2.8), provided $p_1(s) > 0$, that $t_1 - \bar{e} = 0$ for all $s \notin C_k \cup C_l$, that $t_1 - \bar{e} = (y_k'' - e_k)/N\alpha_k$ for all $s \in C_k - C_{kl} = C_k - \{s : k, l \in s\}$, and that $t_1 - \bar{e} = (y_l'' - e_l)/N\alpha_l$ for all $s \in C_l - C_{kl}$. Hence, for $\mathbf{y} = \mathbf{y}''$, the first moment of $t_1 - \bar{e}$ can be written

$$E_{p_1}(t_1 - \bar{e}) = E_{p_1}^{kl}(t_1 - \bar{e})$$

$$-\alpha_{kl}(1)\left\{\frac{y_k'' - e_k}{N\alpha_k} + \frac{y_l'' - e_l}{N\alpha_l}\right\} + \frac{y_k'' - e_k}{N} + \frac{y_l'' - e_l}{N}$$

where, for any $k \neq l, p$, and $t$, the new operator $E_p^{kl}$ is used as follows:

$$E_p^{kl}(t) = \Sigma_{C_{kl}} p(s)t$$

and, for $j = 0$ and $1$,

$$\alpha_{kl}(j) = \Sigma_{C_{kl}} p_j(s)$$

Combining this with the fact that $t_1$ is unbiased gives, for $\mathbf{y} = \mathbf{y}''$,

$$E_{p_1}^{kl}(t_1 - \bar{e}) = \alpha_{kl}(1)\left\{\frac{y_k'' - e_k}{N\alpha_k} + \frac{y_l'' - e_l}{N\alpha_l}\right\} \tag{2.9}$$

Also, from the Cauchy–Schwartz inequality,

$$\{E_{p_1}^{kl}(t_1 - \bar{e})\}^2 \leqslant \alpha_{kl}(1) E_{p_1}^{kl}\{(t_1 - \bar{e})^2\} \tag{2.10}$$

For $\mathbf{y} = \mathbf{y}''$, the second moment of $t_1 - \bar{e}$ can be written

$$E_{p_1}\{(t_1 - \bar{e})^2\} = E_{p_1}^{kl}\{(t_1 - \bar{e})^2\}$$

$$-\alpha_{kl}(1)\left\{\frac{(y_k'' - e_k)^2}{N^2\alpha_k^2} + \frac{(y_l'' - e_l)^2}{N^2\alpha_l^2}\right\} + \frac{(y_k'' - e_k)^2}{N^2\alpha_k} + \frac{(y_l'' - e_l)^2}{N^2\alpha_l}$$

from which it follows, by use of (2.9) and (2.10), that

$$E_{p_1}\{(t_1 - \bar{e})^2\} \geqslant \frac{(y_k'' - e_k)^2}{N^2 \alpha_k} + \frac{(y_l'' - e_l)^2}{N^2 \alpha_l} + \frac{2\alpha_{kl}(1)(y_k'' - e_k)(y_l'' - e_l)}{N^2 \alpha_k \alpha_l}$$

Comparing, for $\mathbf{y} = \mathbf{y}''$, this latter expression with

$$E_{p_0}\{(t_{GD} - \bar{e})^2\} = \frac{(y_k'' - e_k)^2}{N^2 \alpha_k} + \frac{(y_l'' - e_l)^2}{N^2 \alpha_l} + \frac{2\alpha_{kl}(0)(y_k'' - e_k)(y_l'' - e_l)}{N^2 \alpha_k \alpha_l}$$

we find, using (2.2), that

$$\alpha_{kl}(1)(y_k'' - e_k)(y_l'' - e_l) \leqslant \alpha_{kl}(0)(y_k'' - e_k)(y_l'' - e_l)$$

This inequality must hold for both positive and negative values of the product $(y_k'' - e_k)(y_l'' - e_l)$. It follows that $\alpha_{kl}(0) = \alpha_{kl}(1)$. Since $k$ and $l$ were arbitrarily chosen, we conclude that $\alpha_{kl}(0) = \alpha_{kl}(1)$ for $k \neq l = 1, \ldots, N$. □

**Theorem 3.2.** *Any strategy* $(p, t_{GD})$, *where* $\Sigma_{\mathscr{S}} p(s)\nu(s) = n$ *and* $t_{GD} = \Sigma_S (y_k - e_k)/N\alpha_k + \bar{e}$ *with an arbitrary vector* $\mathbf{e} = (e_1, \ldots, e_N)$, *is admissible in* $\mathscr{H}_u(n)$.

**Proof.** Let $(p_0, t_{GD})$ be an arbitrary strategy with $\Sigma_{\mathscr{S}} p_0(s)\nu(s) = n$. Suppose there is a strategy $(p_1, t_1)$ in $\mathscr{H}_u(n)$ such that

$$V(p_1, t_1) \leqslant V(p_0, t_{GD}) \quad \text{for all } \mathbf{y} \in R_N \tag{2.11}$$

Then, from Lemma 3.2, $\alpha_k(0) = \alpha_k(1)$ and $\alpha_{kl}(0) = \alpha_{kl}(1)$ $(k \neq l = 1, \ldots, N)$, and hence $V(p_0, t_{GD}) = V(p_1, t_{GD})$ for all $\mathbf{y} \in R_N$. Since $t_{GD}$ is an admissible estimator in $\mathscr{A}_u$ under the design $p_1$ (see Theorem 3.1), strict inequality cannot hold in (2.11) for any $\mathbf{y} \in R_N$. Hence the strategy $(p_1, t_1)$ cannot be better than $(p_0, t_{GD})$. We conclude that no strategy in $\mathscr{H}_u(n)$ can be better than $(p_0, t_{GD})$, which means that $(p_0, t_{GD})$ is admissible. □

A consequence of Theorem 3.2 is that there does not exist a unique strategy in $\mathscr{H}_u(n)$ that is better than all other strategies in $\mathscr{H}_u(n)$. For there are infinitely many strategies of the type $(p, t_{GD})$ in $\mathscr{H}_u(n)$, corresponding to the infinitely many choices of $p$ and $\mathbf{e}$, and all of them are admissible.

By considering special values of the constant vector $\mathbf{e}$, we obtain various corollaries of Theorem 3.2. For example, when $\mathbf{e} = \mathbf{0}$, $t_{GD}$ is equal to $t_{HT}$, and hence we obtain Ramakrishnan's (1975) result:

**Corollary 3.3.** *Any strategy* $(p, t_{HT})$ *with* $\Sigma_{\mathscr{S}} p(s)\nu(s) = n$, *is admissible in* $\mathscr{H}_u(n)$. □

By considering $\mathbf{e} = c\mathbf{x}$, where $c$ is a predetermined constant, and $\mathbf{x}$ is a vector of known values of an auxiliary variable, we obtain:

**Corollary 3.4.** *Any strategy* $(p, t_{GD})$ *with* $\Sigma_{\mathscr{S}} p(s)\nu(s) = n$, *and* $t_{GD} = \Sigma_S y_k/N\alpha_k + c(\bar{x} - \Sigma_S x_k/N\alpha_k)$, *is admissible in* $\mathscr{H}_u(n)$. □

**Remark 1.** It follows from Corollaries 3.3 and 3.4 that both of the FES($n$) strategies $(srs, \bar{y}_S)$ and $(srs, t_D)$ with $t_D = \bar{y}_S + c(\bar{x} - \bar{x}_S)$ are admissible strategies in $\mathscr{H}_u(n)$. ■

**Remark 2.** Assume that we consider strategies in $\mathscr{H}_u(n)$. Since the strategy $(srs, \bar{y}_S)$ is admissible, it cannot be improved upon, for example, by going to a PPS-sampling strategy such as the Horvitz–Thompson strategy $(ppsx, \bar{x}R_{yx})$, where $\bar{x}R_{yx}$ is given by (6.11) of Section 1.6. Compare also Remarks 3 and 9 of Section 3.1.

On the other hand, $(ppsx, \bar{x}R_{yx})$ is also admissible and likewise impossible to improve upon within $\mathscr{H}_u(n)$. This illustrates the weakness of the admissibility criterion in selection of a universally good strategy. ■

Finally, let us drop the condition of unbiasedness and look for admissible strategies in $\mathscr{H}(n)$, the class of all (biased and unbiased) strategies with expected effective sample size $n$. We state without proof the following theorem, due to Joshi (1966):

**Theorem 3.3.** *Any strategy* $(p, \bar{y}_S)$, *where* $p$ *is an FES(n) design, is admissible in* $\mathscr{H}(n)$. □

**Remark 3.** It follows from Theorem 3.3 that the FES($n$) strategy $(srs, \bar{y}_S)$ is admissible in $\mathscr{H}(n)$. ■

**Remark 4.** There are many FES($n$) designs $p$ such that the strategy $(p, \bar{y}_S)$ is biased, but according to Theorem 3.3 these strategies are still admissible in $\mathscr{H}(n)$. Thus if we adopt the admissibility criterion biased strategies $(p, \bar{y}_S)$ are not excluded from consideration, as long as $n < N$. (When $n = N$, all biased strategies in $\mathscr{H}(n)$ are easily seen to be inadmissible.) ■

## 3.3. HYPERADMISSIBILITY

Having found in Section 3.1 that the criterion of admissibility is not sharp enough to produce a unique estimator, we now consider the criterion of *hyperadmissibility*, a stronger form of admissibility. This new criterion essentially singles out one estimator, the Horvitz–Thompson estimator, as the unique hyperadmissible estimator in the class $\mathscr{L}_{0u}$, and also in the class $\mathscr{A}_u$.

The concept of hyperadmissibility is due to Hanurav (1965, 1968), who gives two slightly different definitions, of which we adopt the one in Hanurav (1965).

The relation between the two definitions is discussed in some detail in Lanke and Ramakrishnan (1974). Hyperadmissibility is also discussed in Joshi (1971, 1972).

The definition of hyperadmissibility requires the introduction of a special type of subset of the parameter space $R_N$ called "principal hypersurface."

**Definition.** The *principal hypersurface* (phs) $\Omega(k_1, \ldots, k_h)$ is the set of all $\mathbf{y} \in R_N$ such that $y_k = 0$ for $k \neq k_1, \ldots, k_h$, and $-\infty < y_k < +\infty$ for $k = k_1, \ldots, k_h$. ■

Thus, a phs is a subset of $R_N$, where certain prescribed coordinates are held equal to zero. There is one phs, $\Omega(k_1, \ldots, k_h)$, for each non-empty subset $\{k_1, \ldots, k_h\}$ of $\{1, \ldots, N\}$ $(h = 1, \ldots, N)$. In all there are $2^N - 1$ phs's. For $h = N$ there is exactly one phs, namely, $\Omega(1, \ldots, N) = R_N$. We note that the zero vector, $\mathbf{0} = (0, \ldots, 0)$, is contained in each phs.

**Definition.** An estimator $t$, belonging to some class $\mathscr{C}$ of estimators, is said to be *hyperadmissible* in $\mathscr{C}$ under a given design $p$, if and only if, for any phs $\Omega(k_1, \ldots, k_h)$ in $R_N$, $t$ is admissible in $\mathscr{C}$ when the parameter space is restricted to $\Omega(k_1, \ldots, k_h)$. ■

It follows immediately from the definition that the stronger property of hyperadmissibility implies admissibility.

Is there any practically motivated rationale behind hyperadmissibility? Hanurav (1968) gives the following argument. Let $y$ denote the population total, $y = \Sigma_1^N y_k$. Let $\mathscr{U}^*$ be a subpopulation, that is a subset of the population $\mathscr{U} = \{1, \ldots, N\}$, with subpopulation total $y^* = \Sigma_{k \in \mathscr{U}^*} y_k$. If we define a new population vector $\mathbf{y}^* = (y_1^*, \ldots, y_N^*)$, where

$$y_k^* = \begin{cases} y_k & \text{for } k \in \mathscr{U}^* \\ 0 & \text{for } k \notin \mathscr{U}^* \end{cases}$$

then $y^*$ can be expressed as a population total with respect to $\mathbf{y}^*$, namely, $y^* = \Sigma_1^N y_k^*$.

Suppose that we have decided, under a given design, to estimate the population total $y$ by a certain estimator $t = t(D)$, where $D = \{(k, y_k)\colon k \in S\}$. Suppose also that, under the same given design, $y^*$ is estimated by $t^* = t(D^*)$, where $D^* = \{(k, y_k^*)\colon k \in S\}$. Since it is frequent practice to estimate $y^*$ in this way, it is desirable that $t^*$ be admissible for $y^*$. From the definition of hyperadmissibility it follows that $t$ is hyperadmissible for estimating the population total $y$ if and only if $t^*$ is admissible for the totals $y^*$ of *all possible* subpopulations $\mathscr{U}^* \subseteq \mathscr{U}$.

The hyperadmissibility criterion has sometimes been deemed to be of limited interest and, in any case, it can not be said to give a conclusive answer to the question for which the criterion was designed, namely: Is there a unique optimal estimator of the finite population mean? One criticism is the following (see, for example, Rao and Singh, 1973): The fact that hyperadmissibility is equivalent to admissibility for all possible subpopulations (as seen above) is of questionable value, because one is seldom interested in all possible subpopulations, and especially not in subpopulations with only one member. Another criticism (see, for example, J. N. K. Rao, 1975) maintains that the very structure of the hyperadmissibility criterion strongly favors the Horvitz–Thompson estimator, among the unbiased estimators. If we were to strictly follow such a criterion, we should use $t_{HT}$ for practically all designs; see Theorem 3.6 below. But it is well known that $t_{HT}$ can perform very badly, for example, from a variance point of view, when the $y_k$'s are poorly or negatively correlated with the $\alpha_k$'s that characterize the given design. Thus, it may be unwise to use $t_{HT}$, unless one knows that $\alpha_k \propto y_k$ approximately.

In our discussion of results on hyperadmissibility, we shall first prove that the Horvitz–Thompson estimator is hyperadmissible in the class $\mathcal{A}_u$. ■

**Theorem 3.4.** *For any given design with* $\alpha_k > 0$ $(k = 1, \ldots, N)$, *the Horvitz–Thompson estimator* $t_{HT} = \Sigma_s y_k/N\alpha_k$ *is hyperadmissible in the class* $\mathcal{A}_u$ *of all unbiased estimators of* $\bar{y}$.

**Proof.** We know from Corollary 3.1, Section 3.1, that $t_{HT}$ is admissible, in $\mathcal{A}_u$, when the parameter space is $R_N$. We also noted that $t_{HT}$ is admissible, even when the parameter space is restricted to a subset of $R_N$, as long as the zero vector, $\mathbf{0} = (0, \ldots, 0)$, belongs to that subset. Since $\mathbf{0}$ is contained in every phs, we conclude that $t_{HT}$ is admissible in $\mathcal{A}_u$, when the parameter space is restricted to any phs in $R_N$. Thus $t_{HT}$ is hyperadmissible in $\mathcal{A}_u$. □

**Remark 1.** It follows immediately that $t_{HT}$ is also hyperadmissible in the class $\mathcal{L}_{0u}$ of all unbiased linear homogeneous estimators of $\bar{y}$. ■

At this point the question arises as to whether there are any other hyperadmissible estimators in $\mathcal{L}_{0u}$ and $\mathcal{A}_u$ besides $t_{HT}$. The answer is given in Theorems 3.5 and 3.6 below. Theorem 3.6 says that $t_{HT}$ is uniquely hyperadmissible in $\mathcal{A}_u$, except for certain very special designs.

In the proof of Theorem 3.5 we shall use the following lemma, which says that $t_{HT}$ is the uniformly minimum variance estimator in $\mathcal{L}_{0u}$, provided that the parameter space is restricted to an arbitrary one-dimensional phs $\Omega(k_0)$ in $R_N$.

**Lemma 3.3.** *For any given design with* $\alpha_k > 0$ $(k = 1, \ldots, N)$, *for any one-dimensional phs* $\Omega(k_0)$, *and for any estimator* $t = \Sigma_s w_{ks} y_k \in \mathcal{L}_{0u}$,

$$V(t_{HT}) \leq V(t) \quad \text{for all } \mathbf{y} \in \Omega(k_0)$$

*with equality if and only if* $w_{k_0 s} = 1/N\alpha_{k_0}$ *for all samples* $s$ *such that* $k_0 \in s$ *and* $p(s) > 0$.

**Proof.** Let $k_0$ be fixed. We shall prove that $E(t_{HT}^2) \leqslant E(t^2)$ when $\mathbf{y} \in \Omega(k_0)$, for any estimator $t \in \mathscr{L}_{0u}$. We note first that

$$E(t_{HT}^2) = \frac{y_{k_0}^2}{N^2 \alpha_{k_0}} \tag{3.1}$$

for all $\mathbf{y} \in \Omega(k_0)$ and

$$E(t^2) = \{\Sigma_{\mathscr{S}_0} p(s)\, w_{k_0 s}^2\}\, y_{k_0}^2 \tag{3.2}$$

for all $y \in \Omega(k_0)$, where $\mathscr{S}_0 = \{s \in \mathscr{S} : k_0 \in s \wedge p(s) > 0\}$. Using the Cauchy–Schwarz inequality we obtain

$$\{\Sigma_{\mathscr{S}_0} p(s)\, w_{k_0 s}\}^2 \leqslant \{\Sigma_{\mathscr{S}_0} p(s)\} \{\Sigma_{\mathscr{S}_0} p(s)\, w_{k_0 s}^2\} \tag{3.3}$$

Since $\Sigma_{\mathscr{S}_0} p(s) w_{k_0 s} = 1/N$ due to the unbiasedness of $t$, and since $\Sigma_{\mathscr{S}_0} p(s) = \alpha_{k_0}$, we have from (3.3) that

$$\Sigma_{\mathscr{S}_0} p(s)\, w_{k_0 s}^2 \geqslant \frac{1}{N^2 \alpha_{k_0}} \tag{3.4}$$

with equality if and only if $w_{k_0 s}$ is equal to some constant $c$ for all $s \in \mathscr{S}_0$ such that $p(s) > 0$. Since $t$ is unbiased, we must have $c = 1/N\alpha_{k_0}$. From (3.1), (3.2), and (3.4) we conclude that $E(t_{HT}^2) \leqslant E(t^2)$ for all $\mathbf{y} \in \Omega(k_0)$ with equality if and only if $w_{k_0 s} = 1/N\alpha_{k_0}$ for all $s \in \mathscr{S}_0$ such that $p(s) > 0$. Thus, the lemma is proved. □

**Theorem 3.5.** *For any given design with* $\alpha_k > 0$ $(k = 1, \ldots, N)$, *the Horvitz–Thompson estimator* $t_{HT} = \Sigma_S y_k / N\alpha_k$ *is the* unique hyperadmissible *estimator in the class* $\mathscr{L}_{0u}$ *of all unbiased linear homogeneous estimators of* $\bar{y}$.

**Proof.** By Remark 1, $t_{HT}$ is hyperadmissible in $\mathscr{L}_{0u}$. We shall now show that $t_{HT}$ is the only member of $\mathscr{L}_{0u}$ that is admissible in certain particular phs's, namely in every phs of dimension one. Once this has been shown, it follows immediately that $t_{HT}$ is the unique hyperadmissible estimator in $\mathscr{L}_{0u}$.

From Lemma 3.3 above it follows that an estimator $t \in \mathscr{L}_{0u}$ is admissible for each one-dimensional phs $\Omega(k)$ in $R_N$ if and only if

$$V(t) = V(t_{HT}) \qquad \text{for all } \mathbf{y} \in \Omega(k)\ (k = 1, \ldots, N)$$

Another consequence of Lemma 3.3 is that these variance equalities hold if and only if, for $k = 1, \ldots, N$, $w_{ks} = 1/N\alpha_k$ for all samples $s$ such that $k \in s$ and $p(s) > 0$, that is, if and only if $P(t = t_{HT}; \mathbf{y}) = 1$ for all $\mathbf{y} \in R_N$. □

We conclude this section by presenting (without proof) a theorem due to Lanke and Ramakrishnan (1974), which gives a complete description of all hyperadmissible estimators in $\mathscr{A}_u$. First we define the term "unicluster design," which is used in the theorem.

**Definition.** A design $p$ is said to be a *unicluster design* if all samples with positive probability are pairwise disjoint, that is, $s_1 \cap s_2 = \emptyset$ (= the empty set) for any pair of samples $s_1$ and $s_2$ such that $p(s_1) > 0, p(s_2) > 0$ and $s_1 \neq s_2$. ■

**Remark 2.** The only unicluster design of practical importance seems to be systematic sampling (with only one starting point); see Cochran (1963). ■

**Theorem 3.6.** *For any given design $p$ with $\alpha_k > 0$ $(k = 1, \ldots, N)$ the following holds:*

*(a) If $p$ is a non-unicluster design or a unicluster design with at least three clusters, then $t_{HT}$ is the unique hyperadmissible estimator in the class $\mathscr{A}_u$ of all unbiased estimators of $\bar{y}$;*
*(b) If $p$ is a unicluster design with exactly two clusters, then every estimator $t \in \mathscr{A}_u$ is hyperadmissible;*
*(c) If $p$ is a unicluster design with exactly one cluster (that is, a census), then $t_{HT}$ is the only member of $\mathscr{A}_u$, and hence, trivially, the unique hyperadmissable estimator in $\mathscr{A}_u$.* □

Thus, for all designs of practical interest, the Horvitz–Thompson estimator is the unique hyperadmissible estimator in $\mathscr{A}_u$.

**Remark 3.** Theorem 3.6 has the following consequence: For any non-unicluster design and any unicluster design with $\alpha_k > 0$ $(k = 1, \ldots, N)$, the generalized difference estimator, $t_{GD} = \Sigma_S(y_k - e_k)/N\alpha_k + \bar{e}$, is not hyperadmissible in $\mathscr{A}_u$, and hence not hyperadmissible in $\mathscr{L}_u \subset \mathscr{A}_u$ (provided, of course, that $\mathbf{e} \neq \mathbf{0}$, so that $t_{GD}$ is not identical to $t_{HT}$). ■

**Remark 4.** Although not hyperadmissible, $t_{GD}$ has a property that resembles hyperadmissibility: For any given design with $\alpha_k > 0$ $(k = 1, \ldots, N)$, and for any given constant vector $\mathbf{e} \in R_N$, $t_{GD}$ is admissible, in $\mathscr{A}_u$ and in $\mathscr{L}_u$, when the parameter space is restricted to *any* subset $\Omega'(k_1, \ldots, k_h)$ of $R_N$, where

$$\Omega'(k_1, \ldots, k_h) = \{\mathbf{y} \in R_N : y_k = e_k \quad \text{for} \quad k \neq k_1, \ldots, k_h,$$
$$\text{and} \quad -\infty < y_k < +\infty \quad \text{for} \quad k = k_1, \ldots, k_h\}$$

This property of $t_{GD}$ follows from the fact that the vector $\mathbf{e}$ belongs to each of the $(2^N - 1)$ possible subsets $\Omega'(k_1, \ldots, k_h)$, and $t_{GD}$ was seen (in Remark 5 of Section 3.1) to be admissible for any parameter space that contains $\mathbf{e}$.

Hence $t_{HT}$ is hyperadmissible while $t_{GD}$ is not. In order to evaluate the significance of this conclusion, recall that hyperadmissibility requires admissibility for any of the subsets of $R_N$ called a phs. Thus, hyperadmissibility is a criterion intimately tied to a very particular kind of subset of $R_N$.

On the other hand, $t_{GD}$ has the property of being admissible for any subset

$\Omega'(k_1, \ldots, k_h)$ defined in this remark; this constitutes what may be called a "generalized hyperadmissibility" property.

By and large, the estimator $t_{GD}$ has just as many appealing features as its special case, $t_{HT}$. Therefore, the lack of hyperadmissibility of $t_{GD}$ emphasizes the feeling that hyperadmissibility is a criterion narrowly tailored to single out $t_{HT}$ as a unique optimal estimator. ■

## 3.4. UNIFORMLY MINIMUM VARIANCE: NONEXISTENCE THEOREMS

In this section and the following one we deal with the existence and nonexistence of uniformly minimum variance (UMV) estimators. In discussing UMV estimators, we consider only unbiased estimators. It is assumed, throughout, that the parameter space equals $R_N$.

**Definition.** An estimator $t_0$, belonging to some class $\mathscr{C}$ of unbiased estimators, is said to be a *uniformly minimum variance* (UMV) estimator in $\mathscr{C}$ under a given design $p$, if and only if $V(t_0) \leqslant V(t)$ for all estimators $t \in \mathscr{C}$, and for all $\mathbf{y} \in R_N$. ■

The term "uniformly" in "uniformly minimum variance" means "for all $\mathbf{y} \in R_N$."

UMV is a very strong property to be required from an estimator. If a UMV estimator existed in some class $\mathscr{C}$, then this estimator would be a natural choice from that class of estimators. However, as will be shown in this section, no UMV estimator exists in the comprehensive class $\mathscr{A}_u$ of all unbiased estimators of $\bar{y}$, nor does one exist in the somewhat more restricted classes $\mathscr{L}_u$ and, under almost all designs, $\mathscr{L}_{0u}$. (In Section 3.5 we shall establish the existence of UMV estimators under more restricted conditions.)

The nonexistence of a UMV estimator in $\mathscr{L}_{0u}$ was first proved by Godambe (1955), and the nonexistence in $\mathscr{A}_u$ by Godambe and Joshi (1965). As a consequence of these nonexistence theorems, the interest was then directed towards other optimality criteria than UMV, for example, admissibility and hyperadmissibility. The additional criteria that have been proposed have not gained widespread acceptance. Those include linear sufficiency, mentioned in Section 2.2.

The following lemma giving a necessary and sufficient condition for the existence of an unbiased estimator of $\bar{y}$ will be the starting point for our discussion of problems associated with UMV estimation.

**Lemma 3.4.** *For any given design $p$ with inclusion probabilities $\alpha_k$ $(k = 1, \ldots, N)$, there is at least one unbiased estimator of $\bar{y}$ if and only if $\alpha_k > 0$ $(k = 1, \ldots, N)$.*

**Proof.** If $\alpha_k > 0$ $(k = 1, \ldots, N)$, then the Horvitz–Thompson estimator $t_{HT} = \Sigma_S y_k/N\alpha_k$ always exists, and it is unbiased for $\bar{y}$. Now suppose that at least one $\alpha_k$ is zero, say $\alpha_{k_0} = 0$. Then the expectation of any estimator $t, E(t) = \Sigma_{\mathscr{S}} p(s)t(d)$, cannot depend on $y_{k_0}$. Hence, $E(t)$ cannot be equal to $\bar{y}$ for all $\mathbf{y} \in R_N$. Thus no unbiased estimator of $\bar{y}$ can exist. □

In searching for UMV estimators in $\mathscr{A}_u$, $\mathscr{L}_u$ and $\mathscr{L}_{0u}$ under various designs, we can disregard all designs with $\alpha_k = 0$ for at least one $k$, because Lemma 3.4 tells us that $\mathscr{A}_u$, $\mathscr{L}_u$, and $\mathscr{L}_{0u}$ are empty under all such designs. Likewise, we can disregard all "census designs," that is, designs with $\alpha_k = 1$ for $k = 1, \ldots, N$, because for any such design $t_{HT}$ is the *only* member of $\mathscr{A}_u$, $\mathscr{L}_u$, and $\mathscr{L}_{0u}$, and hence, trivially, UMV in those classes. Thus, in the rest of this section, we limit consideration to noncensus designs with $\alpha_k > 0$ $(k = 1, \ldots, N)$.

The following theorem, which establishes the nonexistence of a UMV estimator in $\mathscr{A}_u$, is due to Godambe and Joshi (1965), but the proof given here is from Basu (1971).

**Theorem 3.7.** *Let any noncensus design with* $\alpha_k > 0$ $(k = 1, \ldots, N)$ *be given. Then no UMV estimator exists in the class* $\mathscr{A}_u$ *of all unbiased estimators of* $\bar{y}$.

**Proof.** For any given parameter vector $\mathbf{y}_0 \in R_N$ there exists an estimator in $\mathscr{A}_u$ with zero variance for $\mathbf{y} = \mathbf{y}_0$, namely, the generalized difference estimator $t_{GD} = \Sigma_S(y_k - e_k)/N\alpha_k + \bar{e}$, with the constant vector $\mathbf{e}$ chosen equal to $\mathbf{y}_0$.

Thus, an estimator $t \in \mathscr{A}_u$ can be UMV in $\mathscr{A}_u$ only if $V(t) = 0$ for *all* $\mathbf{y} \in R_N$, which is possible only in a census. Hence, no UMV estimator of $\bar{y}$ exists in $\mathscr{A}_u$. □

**Remark 1.** Since $t_{GD} \in \mathscr{L}_u$ for all constant vectors $\mathbf{e} \in R_N$, we can prove the following result in the same way as Theorem 3.7 above: Consider any given noncensus design with $\alpha_k > 0$ $(k = 1, \ldots, N)$, then no UMV estimator exists in the class $\mathscr{L}_u$ of all unbiased linear (homogeneous and nonhomogeneous) estimators of $\bar{y}$. ■

Thus, the nonexistence of a UMV estimator applies within the class $\mathscr{A}_u$ as well as within the more restricted class $\mathscr{L}_u$. In the next theorem we shall see that in the even further restricted class $\mathscr{L}_{0u}$, there is still no UMV estimator, under practically all designs. This theorem was originally shown by Godambe (1955). The first entirely satisfactory proof seems to have been given by Lanke (1973, 1975), who also presents a detailed discussion of earlier proofs. Our proof below, including the preceding lemma, follows essentially the line of reasoning in Lanke (1973, 1975).

**Lemma 3.5.** *Let any non-unicluster design with $\alpha_k > 0$ $(k = 1, \ldots, N)$ be given. Then the Horvitz–Thompson estimator $t_{HT} = \Sigma_s y_k / N\alpha_k$ is not UMV in the class $\mathscr{L}_{0u}$ of all unbiased linear homogeneous estimators of $\bar{y}$.*

**Proof.** Consider an arbitrary non-unicluster design with $\alpha_k > 0$ $(k = 1, \ldots, N)$. Then there must be at least two nondisjoint and nonidentical samples, say $s_1$ and $s_2$, with $p(s_1) > 0$ and $p(s_2) > 0$. Suppose for simplicity that the label $k = 1$ appears in both $s_1$ and $s_2$, and the label $k = 2$ in $s_1$ but not in $s_2$. Let the estimator $t_0 \in \mathscr{L}_{0u}$ be defined by

$$t_0(d) = \begin{cases} t_{HT}(d) - p(s_2)y_1 & \text{for } s = s_1 \\ t_{HT}(d) + p(s_1)y_1 & \text{for } s = s_2 \\ t_{HT}(d) & \text{for } s \neq s_1, s_2 \end{cases}$$

Consider, for example, $\mathbf{y} = (a, b, 0, \ldots, 0)$. We then have

$$V(t_{HT}) - V(t_0) = p(s_1)p(s_2)a[2b/N\alpha_2 - a\{p(s_1) + p(s_2)\}] \tag{4.1}$$

From (4.1) we see that $V(t_0)$ can be smaller than $V(t_{HT})$ for certain values of $a$ and $b$; for example, when $a > 0$ and $b > N\alpha_2 a\{p(s_1) + p(s_2)\}/2$. Hence $t_{HT}$ is not UMV in $\mathscr{L}_{0u}$. □

**Theorem 3.8.** *Let any noncensus design $p$ with $\alpha_k > 0$ $(k = 1, \ldots, N)$ be given. Then the following holds:*

*(a) If $p$ is a non-unicluster design, no UMV estimator exists in the class $\mathscr{L}_{0u}$ of all unbiased linear homogeneous estimators of $\bar{y}$;*

*(b) If $p$ is a unicluster design, the Horvitz–Thompson estimator $t_{HT} = \Sigma_s y_k / N\alpha_k$ is the only member of $\mathscr{L}_{0u}$, and hence, trivially, UMV in $\mathscr{L}_{0u}$.*

**Proof.** (a) From Lemma 3.5 it follows that an estimator $t \in \mathscr{L}_{0u}$ can be UMV in $\mathscr{L}_{0u}$ only if

$$V(t) \leqslant V(t_{HT}) \tag{4.2}$$

for all $\mathbf{y} \in R_N$, with strict inequality for at least one $\mathbf{y} \in R_N$. But since $t_{HT}$ is admissible in $\mathscr{L}_{0u}$ (see Remark 4 in Section 3.1), no estimator in $\mathscr{L}_{0u}$ can satisfy (4.2). Hence, no UMV estimator exists in $\mathscr{L}_{0u}$.

(b) It is known that a linear homogeneous estimator $t = \Sigma_s w_{ks} y_k$ is unbiased for $\bar{y}$ if and only if

$$\Sigma_{C_k} p(s)\, w_{ks} = \frac{1}{N} \quad (k = 1, \ldots, N) \tag{4.3}$$

Under a unicluster design, each label $k$ belongs to exactly one sample. Denoting by $s_k$ the sample to which the label $k$ belongs, we can write the necessary and sufficient condition for unbiasedness (4.3) as $p(s_k)w_{ks_k} = 1/N$ $(k = 1, \ldots, N)$. But under a unicluster design $p(s_k) = \alpha_k$; hence the condition is satisfied if and only if $w_{ks_k} = 1/N\alpha_k$ $(k = 1, \ldots, N)$. Therefore, $t_{HT}$ is the *only* linear homogeneous estimator that is unbiased for $\bar{y}$ under a unicluster design. □

**Remark 2.** In the proof of Lemma 3.5, we encountered one estimator $t_0$ such that $V(t_0) < V(t_{HT})$ for certain values of $\mathbf{y}$. We concluded from this that $t_{HT}$ is not UMV in $\mathscr{L}_{0u}$, and, consequently, not in $\mathscr{A}_u$. An additional example is Example 1 in Section 1.5 where it was shown that, under *srs*, the sample mean $\bar{y}_S$ is neither UMV in $\mathscr{L}_{0u}$ nor UMV in $\mathscr{A}_u$. ■

**Remark 3.** So far we have discussed the UMV property of *estimators* under a given design. It is also possible to talk about UMV *strategies*, defined as follows: A strategy $(p_0, t_0)$, belonging to some class $\mathscr{H}$ of unbiased strategies, is said to be a UMV strategy in $\mathscr{H}$, if and only if $V(p_0, t_0) \leqslant V(p, t)$ for all strategies $(p, t) \in \mathscr{H}$, and for all $\mathbf{y} \in R_N$. It follows from Theorem 3.7 (and also from Theorem 3.2) that no UMV strategy exists in the class $\mathscr{H}_u(n)$ of all unbiased strategies with expected effective size equal to $n$, provided that $n < N$. Nonexistence of UMV strategies in other classes of strategies can also easily be established. ■

## 3.5. UNIFORMLY MINIMUM VARIANCE: EXISTENCE THEOREMS UNDER RESTRICTED CONDITIONS

Although no UMV estimators exist in the classes $\mathscr{A}_u$, $\mathscr{L}_u$, and $\mathscr{L}_{0u}$ (for practically all designs), we shall now see that UMV estimators indeed exist in more restricted classes of estimators, and under certain designs. The main theorem of the section shows that the generalized difference estimator is UMV under certain conditions. Among the various corollaries of this theorem we recognize results obtained earlier by Neyman (1934), Royall (1968), Särndal (1972, 1976), and Watson (1964).

Throughout this section we consider ordered designs, satisfying the following three conditions:

(i) $p(\mathbf{s}) > 0 \Rightarrow \mathbf{s}$ is a sequence of $n$ distinct labels from $\{1, 2, \ldots, N\}$, that is, the design is both FS($n$) and FES($n$),
(ii) All $n!$ sequences, $\mathbf{s}$, that are permutations of the same labels have the same probability,
(iii) $\alpha_k > 0$ for $k = 1, \ldots, N$.

Suppose that $\mathbf{e} = (e_1, \ldots, e_N)$ is a known, predetermined constant vector in

$R_N$. For given $\mathbf{e} \in R_N$, and a given design, we define

$$z_k = \frac{n(y_k - e_k)}{N\alpha_k} \quad (k = 1, \ldots, N) \tag{5.1}$$

Under the type of design described in (i)–(iii) above, we obtain a sample $\mathbf{S} = (K_1, \ldots, K_n)$ and observe the data $\mathbf{D} = ((k, y_k): k \in \mathbf{S}) = ((K_1, y_{K_1}), \ldots, (K_n, y_{K_n}))$. We shall consider, in this section, the class of unbiased estimators of $\bar{y}$ that depend on $\mathbf{D}$ only through the sequence of unlabeled quantities $(z_{K_1}, \ldots, z_{K_n})$, where $z_{K_i} = n(y_{K_i} - e_{K_i})/N\alpha_{K_i}$ $(i = 1, \ldots, n)$.

**Remark 1.** The estimators in the class described above are not totally independent of the labels, but they depend on labels only in a restricted way. Information about the sampled labels is utilized in order to attach the appropriate values of $e_k$ and $\alpha_k$ to the observed $y_k$-values. But once the transformed $y_k$-values $z_k = n(y_k - e_k)/N\alpha_k$ have been derived, for $k \in$ s, no further use of the labels is allowed, the implicit assumption being that the labels have no more information to give. ■

For the proof of Theorem 3.9 below we need two lemmas, the first of which deals with exchangeable random variables $X_1, \ldots, X_n$ without specific reference to survey sampling. Exchangeability, which will be further discussed in Chapters 4 and 5, is defined as follows: The random variables $X_1, \ldots, X_n$ are said to be *exchangeable* if all $n!$ permutations $(X_{i_1}, \ldots, X_{i_n})$ have the same $n$-dimensional probability distribution.

**Lemma 3.6.** *Let $X_1, \ldots, X_n$ be exchangeable random variables with common expectation $\mu = E(X_i)$ $(i = 1, \ldots, n)$. Then a minimum variance estimator of $\mu$, based on $(X_1, \ldots, X_n)$, must necessarily be a symmetric function of $(X_1, \ldots, X_n)$, that is, it must have the property that $t(X_{i_1}, \ldots, X_{i_n}) = t(X_1, \ldots, X_n)$ for all $n!$ permutations $(X_{i_1}, \ldots, X_{i_n})$.*

**Proof.** Let $t = t(X_1, \ldots, X_n)$ be an arbitrary unbiased, not necessarily symmetric, estimator of $\mu$ with $V(t) = \sigma^2$. Let $\pi = (i_1, \ldots, i_n)$ be an arbitrary permutation of $(1, \ldots, n)$. Consider $t_\pi = t(X_{i_1}, \ldots, X_{i_n})$.

From the definition of exchangeability we know that $(X_{i_1}, \ldots, X_{i_n})$ has the same distribution as $(X_1, \ldots, X_n)$. Hence, $t_\pi$ has the same distribution as $t$, and, in particular, $E(t_\pi) = E(t) = \mu$ and $V(t_\pi) = V(t) = \sigma^2$. A symmetric estimator $t_0$ can be constructed as

$$t_0 = \sum_\pi \frac{t_\pi}{n!}$$

where $\Sigma_\pi$ is over all $n!$ permutations $\pi$ of $(1, \ldots, n)$. Clearly $t_0$ is unbiased for

$\bar{y}$, and

$$V(t_0) = \frac{\sum_{\pi} V(t_\pi) + \sum\sum_{\pi \neq \pi'} [\{V(t_\pi)V(t_{\pi'})\}^{1/2} \rho(t_\pi, t_{\pi'})]}{(n!)^2}$$

$$\leqslant \frac{n!\sigma^2 + n!(n! - 1)\sigma^2}{(n!)^2} = \sigma^2 = V(t)$$

with equality if and only if $t$ is symmetric to start with.

Thus, for any nonsymmetric unbiased estimator $t$ of $\mu$, we can find a symmetric unbiased estimator $t_0$ such that $V(t_0) < V(t)$, and the lemma is proved. □

In the next lemma we return to the survey sampling situation: Consider an FES($n$) design, not necessarily ordered, with $\alpha_k > 0$ ($k = 1, \ldots, N$). Let as usual $S$ be the set of $n$ distinct labels in the sample, and let $z_S = \{z_k: k \in S\}$, where $z_k$ is given by (5.1). Note that $z_S$ is a set of *n not necessarily distinct* numbers (see Remark 1 in Section 1.5). The lemma, which generalizes a result by Royall (1968), says that $z_S$ is a complete statistic (cf. Section 2.3).

**Lemma 3.7.** *Let $p$ be an arbitrarily given FES(n) design with $\alpha_k > 0$ ($k = 1, \ldots, N$). If $g(z_S)$ is a real-valued statistic such that $E\{g(z_S)\} = 0$ for all $\mathbf{y} \in R_N$, then $g(z_s) = 0$ for all $s \in \mathscr{S}$, and for all $\mathbf{y} \in R_N$.*

**Proof.** Let $g(z_S)$ be an arbitrary function of $z_S$, such that $E\{g(z_S)\} = 0$ for all $\mathbf{y} \in R_N$. We shall see that $g(z_S) = 0$ for all $s \in \mathscr{S}$, and for all $\mathbf{y} \in R_N$. Assume for simplicity that $p(s) > 0$ for $s = \{1, 2, \ldots, n\}$. Then it follows that $\alpha_1 > 0$, $\alpha_{12} > 0, \ldots, \alpha_{12\ldots n} > 0$.

First, consider $\mathbf{y} = \mathbf{e}$. Then $z_s = \{0, 0, \ldots, 0\}$ for all $s \in \mathscr{S}$, and $g(z_s)$ is constant. Since $E\{g(z_S)\} = 0$, this constant can be nothing but zero. Thus $g(\{0, 0, \ldots, 0\}) = 0$, that is $g(z_s) = 0$ for all $s \in \mathscr{S}$, when $\mathbf{y} = \mathbf{e}$.

Next, consider $\mathbf{y} = (y_1, e_2, \ldots, e_N)$ for arbitrary $y_1$. Then

$$z_s = \begin{cases} \{z_1, 0, \ldots, 0\} & \text{if } 1 \in s \\ \{0, 0, \ldots, 0\} & \text{otherwise} \end{cases}$$

and

$$E\{g(z_S)\} = \alpha_1 g(\{z_1, 0, \ldots, 0\}) + (1 - \alpha_1) g(\{0, 0, \ldots, 0\})$$

Since $E\{g(z_S)\} = 0$, $\alpha_1 > 0$ and $g(\{0, 0, \ldots, 0\}) = 0$, we conclude that $g(\{z_1, 0, \ldots, 0\}) = 0$. Thus $g(z_s) = 0$ for all $s \in \mathscr{S}$, when $\mathbf{y} = (y_1, e_2, \ldots, e_N)$ for arbitrary $y_1$.

Proceeding similarly we find that $g(z_S) = 0$ for all $s \in \mathscr{S}$, when $\mathbf{y} = (y_1, \ldots,$

$y_l, e_{l+1}, \ldots, e_N$) for arbitrary $y_1, \ldots, y_l$ and for $l = 0, 1, 2, \ldots, N$. When, finally, we reach $l = N$, the lemma is proved. □

Now we can prove the main result of this section.

**Theorem 3.9.** *For any given ordered design satisfying conditions (i)–(iii) above, and for any constant vector* $\mathbf{e} \in R_N$, *the generalized difference estimator*

$$t_{GD} = \Sigma_S \frac{y_k - e_k}{N\alpha_k} + \bar{e} = \bar{z}_S + \bar{e}$$

*is UMV in the class of all unbiased estimators of* $\bar{y}$ *that depend on* **D** *only through the sequence of unlabeled quantities* $(z_{K_1}, \ldots, z_{K_n})$.

**Proof.** It follows from the assumptions about the design that all permutations of the random variables $(z_{K_1}, \ldots, z_{K_n})$ have the same joint distribution with the common expectation $E(z_{K_i}) = \bar{y} - \bar{e}$. Hence we conclude from Lemma 3.6 that a minimum variance unbiased estimator of $\bar{y} - \bar{e}$ in the class of unbiased estimators of $\bar{y} - \bar{e}$ that depend on **D** only through $(z_{K_1}, \ldots, z_{K_n})$ must be symmetric in $(z_{K_1}, \ldots, z_{K_n})$. Thus, it must be a function of $z_S = \{z_k : k \in S\}$, where $S$ is the set of distinct labels that occur in the sequence **S**. Now $\bar{z}_S = \Sigma_S z_k / n$ is a function of $z_S$, and we know that $\bar{z}_S$ is an unbiased estimator of $\bar{y} - \bar{e}$. From Lemma 3.7 it follows that $\bar{z}_S$ is the *only* unbiased estimator of $\bar{y} - \bar{e}$ that is a function of $z_S$. Hence $\bar{z}_S$ is UMV in the class of all unbiased estimators of $\bar{y} - \bar{e}$ that depend on **D** only through $(z_{K_1}, \ldots, z_{K_n})$.

It follows that $t_{GD} = \bar{z}_S + \bar{e}$ is UMV in the class of all unbiased estimators of $\bar{y}$ that depend on **D** only through $(z_{K_1}, \ldots, z_{K_n})$. For if $t$ is an arbitrary function of $(z_{K_1}, \ldots, z_{K_n})$ and unbiased for $\bar{y}$, then

$$V(t) = V(t - \bar{e}) \geqslant V(\bar{z}_S) = V(\bar{z}_S + \bar{e}) = V(t_{GD}). \qquad \square$$

A few corollaries of Theorem 3.9 are of interest. For the first one, let $\mathbf{e} = \mathbf{0}$; this leads to a statement on the Horvitz–Thompson estimator due to Särndal (1976). For the second one, let $\mathbf{e} = \mathbf{0}$ and restrict the design to *srs*; this leads to a statement about the sample mean due to Watson (1964).

**Corollary 3.5.** *For any given ordered design satisfying conditions (i)–(iii) above, the Horvitz–Thompson estimator* $t_{HT} = \Sigma_S y_k / N\alpha_k$ *is UMV in the class of all unbiased estimators of* $\bar{y}$ *that depend on* **D** *only through the sequence of unlabeled quantities* $(n y_{K_1} / N\alpha_{K_1}, \ldots, n y_{K_n} / N\alpha_{K_n})$. □

**Corollary 3.6.** *Under the design srs, the sample mean* $\bar{y}_S = \Sigma_S y_k / n$ *is UMV in the class of all unbiased estimators of* $\bar{y}$ *that depend on* **D** *only through the unlabeled data* $\mathbf{y}_S = (y_{K_1}, \ldots, y_{K_n})$. □

Another consequence we can draw from Theorem 3.9 has to do with the subclass of *linear* estimators that is, the class of estimators $t$ that can be written on the form

$$t = c_0 + \Sigma_1^n c_i z_{K_i}$$

In other words $t$ is linear in $z_{K_i} = n(y_{K_i} - e_{K_i})/N\alpha_{K_i}$, and the coefficients $c_i$ depend on the drawing order $i$ but not on the label $K_i$ appearing in the $i$th draw. The estimator $t_{GD}$, which obtains if $c_0 = \bar{e}$ and $c_i = 1/n$ ($i = 1, \ldots, n$), is a member of the class. Therefore, we also have the following corollary of Theorem 3.9, of which case (c) appeared in Neyman's (1934) pioneering sampling paper. The extension (b) was given in Särndal (1972).

**Corollary 3.7.** *(a) For any given ordered design satisfying conditions (i)–(iii) above, and for any constant vector* $\mathbf{e} \in R_N$*, the generalized difference estimator* $t_{GD}$ *is UMV in the class of all linear unbiased estimators of* $\bar{y}$ *that can be written as*

$$t = c_0 + \Sigma_1^n c_i z_{K_i}$$

*where* $z_{K_i} = n(y_{K_i} - e_{K_i})/N\alpha_{K_i}$.

*(b) Under the same conditions, the Horvitz–Thompson estimator* $t_{HT}$ *is UMV in the class of all linear unbiased estimators of* $\bar{y}$ *that can be written as*

$$t = \Sigma_1^n \frac{c_i n y_{K_i}}{N\alpha_{K_i}}$$

*(c) In particular, under the design srs, the sample mean* $\bar{y}_S$ *is UMV in the class of all linear unbiased estimators of* $\bar{y}$ *that can be written as*

$$t = \Sigma_1^n c_i y_{K_i} \qquad \square$$

Finally, consider estimators that are functions of the unlabeled $z$-data $\{z_k : k \in S\}$, that is, estimators that depend neither on labels nor on drawing order. This case is of interest for unordered designs. The following corollary is a direct consequence of Lemma 3.7; the result (c) is due to Royall (1968).

**Corollary 3.8.** *(a) For any given FES(n) design, and for any constant vector* $\mathbf{e} \in R_N$*, the generalized difference estimator* $t_{GD}$ *is the* only *function of* $\{z_k : k \in S\}$ *that is unbiased for* $\bar{y}$.

*(b) For any given FES(n) design, the Horvitz–Thompson estimator* $t_{HT}$ *is the* only *function of* $\{ny_k/N\alpha_k : k \in S\}$ *that is unbiased for* $\bar{y}$.

*(c) Under the design srs, the sample mean* $\bar{y}_S$ *is the* only *function of* $\{y_k : k \in S\}$ *that is unbiased for* $\bar{y}$. $\square$

**Remark 2.** The essence of this section can be expressed in the following way: By Lemma 3.6, we can neglect information about the drawing order when considering estimators of $\bar{y}$ that are functions of $(z_{K_1}, \ldots, z_{K_n})$. By Lemma 3.7, we are then left with only one estimator, namely, $\bar{z}_S + \bar{e}$, provided the design satisfies the stated conditions. ∎

**Remark 3.** Let us return to the scale-load approach which was discussed in Remark 6 of Section 2.4 only in the special case of the design *srs*. For any given FES($n$) design $p$ with inclusion probabilities $\alpha_k > 0$ ($k = 1, \ldots, N$), define the new scale-points $x_1, \ldots, x_M$ ($M \leqslant N$) as the distinct numbers among $ny_k/N\alpha_k$ ($k = 1, \ldots, N$). The population may be described by the vector $N_1, \ldots, N_M$, where $N_h$ is the multiplicity of $x_h$ in the population. Since the scale points are assumed known, as in Remark 6, Section 2.4, many of the $N_h$ may be zero. For $h = 1, \ldots, M$, let $n_h$ be the number of times that $x_h$ occurs in $\{ny_k/N\alpha_k : k \in s\}$.

Under this scale-load formulation, Särndal (1976) showed that $t_{HT}$ is UMV unbiased estimator (in the case of an ordered FS($n$) and FES($n$) design) and unique unbiased estimator (in the case of an unordered FES($n$) design) among all unbiased estimators that are functions of $n_1, \ldots, n_M$; that is, estimators that ignore the labels.

Thus the scale-load approach provides an alternative proof of the UMV results in Corollary 3.7. Hartley and Rao (1968) used the approach to establish the UMV property of $\bar{y}_s$ for the design *srs*. ■

**Remark 4.** In this section we consider classes of estimators that depend on the labels only in a limited fashion. It would be legitimate to use such estimators in a situation where we cannot gainfully utilize a possible relationship between $k$ and $z_k = n(y_k - e_k)/N\alpha_k$. Intuitively, in such a situation, $k$ is of no value to the inference, once the $z_k$ have been formed.

Lack of relationship between $k$ and $z_k$ is expressed more formally in Chapter 4 through several models, for example, the random permutation model; see Section 4.2. In conformity with the results of this section we shall find in Chapter 4 that $t_{GD}$ and $t_{HT}$ have optimality properties under such models. ■

## 3.6. MINIMAX ESTIMATORS AND STRATEGIES

Several authors, including Aggarwal (1959, 1966), Chaudhuri (1969), Godambe (1960), Godambe and Joshi (1965), Joshi (1966, 1968), Royall (1970b), Scott and Smith (1975), have considered minimax estimation of the finite population mean. We shall center our discussion around some important results concerning the minimax property of procedures involving the sample mean $\bar{y}_s = \Sigma_s y_k/n$.

**Definition.** (a) An estimator $t_0$, belonging to some class $\mathscr{C}$ of estimators, is said to be a *minimax estimator* in $\mathscr{C}$ under a given design $p$ if and only if

$$\max_{\mathbf{y} \in \Omega} \text{MSE}(t_0) \leqslant \max_{\mathbf{y} \in \Omega} \text{MSE}(t)$$

for any estimator $t \in \mathscr{C}$. ($\Omega$ is the parameter space of $\mathbf{y}$.)

(b) A strategy $(p_0, t_0)$, belonging to some class $\mathscr{H}$ of strategies, is said to be a *minimax strategy* in $\mathscr{H}$, if and only if

$$\max_{\mathbf{y}\in\Omega} \mathrm{MSE}(p_0, t_0) \leqslant \max_{\mathbf{y}\in\Omega} \mathrm{MSE}(p, t)$$

for any strategy $(p, t) \in \mathscr{H}$. ■

In the following theorem we shall consider certain subspaces of the parameter space $R_N$: For any given vector $\mathbf{y} \in R_N$ we let $\Omega(\mathbf{y})$ denote the set of all vectors obtained by permutation of the components of $\mathbf{y} = (y_1, \ldots, y_N)$, where $y_1, \ldots, y_N$ are fixed unknown numbers.

**Theorem 3.10.** *Any strategy $(p_o, \bar{y}_S)$, where $p_o$ is an FES(n) design with $\alpha_k = f = n/N$ $(k = 1, \ldots, N)$, is a minimax strategy in the class of all unbiased FES(n) strategies, when the parameter space is restricted to any given subspace $\Omega(\mathbf{y})$, where $\mathbf{y} \in R_N$.*

**Proof.** In proving the theorem, we shall use a result to be presented in Chapter 4, namely, Remark 1 of Section 4.5. That result holds in particular for the random permutation model denoted Model $\mathrm{E}_{RPO}$ (see Section 4.2, especially Table 4.1), under which each of the points in the given subspace $\Omega(\mathbf{y})$ is assigned the probability $1/N!$.

Consider $\Omega(\mathbf{y}_0)$ where $\mathbf{y}_0$ is an arbitrarily given vector in $R_N$. By Remark 1 of Section 4.5, we have, under Model $\mathrm{E}_{RPO}$,

$$\mathscr{E}\, V(p, t) \geqslant \mathscr{E}\, V(p_o, \bar{y}_S)$$

where $\mathscr{E}$ denotes expected value with respect to the distribution that assigns probability $1/N!$ to each point in $\Omega(\mathbf{y}_0)$. The inequality holds for any unbiased FES($n$) strategy $(p, t)$; equality is attained if an only if $(p, t) = (p_o, \bar{y}_S)$. Now, letting $\sigma_y^2 = \Sigma_1^N (y_k - \bar{y})^2/N$, we have

$$V(p_0, \bar{y}_S) = \frac{\{(N-n)/(N-1)\}\sigma_y^2}{n}$$

which remains constant for all $N!$ points in $\Omega(\mathbf{y}_0)$. Thus we must have

$$\max_{\Omega(\mathbf{y}_0)} V(p, t) \geqslant \max_{\Omega(\mathbf{y}_0)} V(p_0, \bar{y}_S) \qquad \square$$

**Remark 1.** The following result follows from Theorem 3.10: For any given FES($n$) design $p$ with $\alpha_k = f$ $(k = 1, \ldots, N)$, $\bar{y}_S$ is a minimax estimator in the class $\mathscr{A}_u$ of all unbiased estimators, when the parameter space is restricted to any given subspace $\Omega(\mathbf{y})$, where $\mathbf{y} \in R_N$. This holds in particular when the given design $p$ is *srs*. ■

**Remark 2.** Among related results, we mention the following: Royall (1970b) showed that $\bar{y}_S$ is minimax estimator among unbiased *linear* estimators, under the design *srs*, when the parameter space is restricted to any given $\Omega(\mathbf{y})$, and the loss function $l(t, \mathbf{y})$ is convex in the first argument.

Aggarwal (1959) showed that $\bar{y}_S$ is minimax estimator (that is, without restriction to unbiasedness or linearity) under the design *srs* if the parameter space is restricted to any given subspace $\{\mathbf{y} : \Sigma_1^N (y_k - \bar{y})^2 \leqslant \text{const.}\}$, and the loss function is quadratic.

Blackwell and Girshick (1954, pp. 229–233) showed that if $t$ is any symmetric estimator, that is, one which depends on $d$ through the unlabeled data $y_s$ only, then the design *srs* is a minimax design in the class of FES($n$) designs, when the parameter space is restricted to any given $\Omega(\mathbf{y})$.

Scott and Smith (1975) considered estimation of the parametric function $\Sigma_1^N M_k y_k$, where the $M_k$ are nonnegative weights and $0 \leqslant y_k \leqslant B$ ($k = 1, \ldots, N$). If $n$ units are sampled with replacement, and if the estimator is $\bar{y}_S \Sigma_1^N M_k$, then the PPS with replacement design that uses probabilities $p_k \propto M_k$ ($k = 1, \ldots, N$) in each draw is under certain conditions an approximately minimax design. ∎

## 3.7. INVARIANT ESTIMATORS

Invariance criteria in connection with estimation of the finite population mean have been discussed notably by Godambe (1968), Godambe and Thompson (1971a), and Basu (1971).

Consider a linear shift of the parameter $\mathbf{y} = (y_1, \ldots, y_N)$ into $\mathbf{y}' = (y'_1, \ldots, y'_N)$, where $y'_k = a + by_k$ ($k = 1, \ldots, N$), for given constants $a, b$; such a shift can be seen as the result of a change of origin and scale of measurement.

The data obtained from any sample $s$ undergo a corresponding shift from $d = \{(k, y_k) : k \in s\}$ into $d' = \{(k, y'_k) : k \in s\}$. Since $d$ and $d'$ represent the same data in two related measurements, the estimate $t(d)$ ought to be related to its value $t(d')$ in the new measurement in a way which agrees formally with the linear shift. The terminology is presented in the following definitions.

**Definition.** The estimator $t(D)$ is said to be *origin invariant* if and only if, for any shift of origin $\mathbf{y} = (y_1, \ldots, y_N)$ into $\mathbf{y}' = (y_1 + a, \ldots, y_N + a)$, it is true that, for all $d$, $t(d') = a + t(d)$. ∎

**Definition.** The estimator $t(D)$ is said to be *scale invariant* if and only if, for any scale shift $\mathbf{y} = (y_1, \ldots, y_N)$ into $\mathbf{y}' = b\mathbf{y} = (by_1, \ldots, by_N)$ such that $b > 0$, it is true that, for all $d$, $t(d') = bt(d)$. ∎

**Definition.** The estimator $t(D)$ is said to be *linear invariant* if and only if, for any origin and scale shift $\mathbf{y} = (y_1, \ldots, y_N)$ into $\mathbf{y}' = (a + by_1, \ldots, a + by_N)$ such that $b > 0$, it is true that, for all $d$, $t(d') = a + bt(d)$. ∎

**Example 1.** The sample mean $\bar{y}_S = \Sigma_S y_k / \nu(S)$ is linear invariant, and hence origin invariant as well as scale invariant.

The difference estimator $t_D = \bar{y}_s + \bar{e} - \bar{e}_S$, is origin invariant, where $\bar{e} = \Sigma_1^N e_k/N$, $\bar{e}_S = \Sigma_S e_k/\nu(S)$, and $\mathbf{e} = (e_1, \ldots, e_N)$ is an arbitrary known vector. (Note that $t_D$ is not scale invariant unless the interpretation of $\mathbf{e}$ is further qualified as in Example 2 below.)

The classical ratio estimator $t_R = \bar{e}\bar{y}_S/\bar{e}_S$ is scale invariant. ■

**Remark 1.** Godambe (1968) has shown that $\bar{y}_S$ is the unique linear invariant estimator in the class of estimators that are functions of the label set $S$ and the sample total $\Sigma_S y_k$. (A justification for limiting consideration to such a class is obtained through the concept of Bayes sufficiency; see Section 6.2.)

Godambe (1968) also showed that the difference estimator $t_D = \bar{y}_S + \bar{e} - \bar{e}_S$ is unique among functions of $(S, \Sigma_S y_k)$ that are origin variant and exactly equal to the population mean $\bar{y}$ when $\mathbf{e} = \mathbf{y} = (y_1, \ldots, y_N)$. Similarly, the ratio estimator $t_R = \bar{e}\bar{y}_S/\bar{e}_S$ is unique among functions of $(S, \Sigma_S y_k)$ that are scale invariant and exactly equal to the population mean $\bar{y}$ when $\mathbf{e} = \mathbf{y}$. ■

Basu (1971) points out that the nonhomogeneous linear estimators, $t(D) = w_{0S} + \Sigma_S w_{kS} y_k$, have been avoided in the belief that such estimators cannot be scale invariant. It is true that a nonhomogeneous linear estimator cannot be scale invariant if $w_{0S}$ does not depend on the scale used. However, the situation becomes different if $w_{0S}$ is considered to be sensitive to changes in the scale of measurement.

**Example 2.** Consider the vector $\mathbf{e}$ to be a preselected point in the parameter space such that, when $\mathbf{y}$ undergoes the linear shift into $\mathbf{y}' = (a + by_1, \ldots, a + by_N)$, $\mathbf{e}$ shifts into $\mathbf{e}' = (a + be_1, \ldots, a + be_N)$. The generalized difference estimator, $t_{GD} = \Sigma_S(y_k - e_k)/N\alpha_k + \bar{e}$ belongs to the class of nonhomogeneous linear estimators with $w_{0S} = \bar{e} - \Sigma_S e_k/N\alpha_k$ and $w_{kS} = 1/N\alpha_k$. We note that $w_{0S}$ is a function of $\mathbf{e}$, so that when $\mathbf{e}$ shifts into $\mathbf{e}'$, the value of $w_{0S}$ will be affected. It is easily seen that $t_{GD}$ is linear invariant, and therefore origin invariant as well as scale invariant.

In particular, the difference estimator $t_D = \bar{y}_S + \bar{e} - \bar{e}_S$ is linear invariant under these conditions. ■

**Remark 2.** In order to be able to compute the linear invariant estimator $t_{GD}$ for the transformed data, one must know beforehand what transformation changed $\mathbf{y}$ into $\mathbf{y}'$ and $\mathbf{e}$ into $\mathbf{e}'$, that is, one must know the constants $a$ and $b$. Thus it would not suffice to know only the parameter point $\mathbf{e}$ prior to the transformation and the data $d'$ after the transformation. This restriction does not apply, for example, to the linear invariant estimator $\bar{y}_S$, the sample mean. The value of $\bar{y}_S$ can be computed as soon as we know the transformed data $d'$. Knowledge of the linear transformation that changed $\mathbf{y}$ into $\mathbf{y}'$ is not required. ■

CHAPTER 4

# Inference under Superpopulation Models: Design-Unbiased Estimation

Chapters 4–6 are devoted to statistical inference in situations where the vector of population values $\mathbf{y} = (y_1, \ldots, y_N)$, is assumed to be the realized outcome of a vector random variable $\mathbf{Y} = (Y_1, \ldots, Y_N)$. The joint distribution of $Y_1, \ldots, Y_N$ will be denoted by $\xi$.

By a "superpopulation model," or simply a "model," we shall mean a specified set of conditions that define a class of distributions to which $\xi$ is assumed to belong. The models to be used are reviewed in Section 4.2.

The salient point of the statistical analysis in the Superpopulation approach is thus that $\mathbf{y}$ is treated as the outcome of $\mathbf{Y}$ with a distribution $\xi$ about which certain features are assumed known.

## 4.1. THE SUPERPOPULATION CONCEPT

Different opinions exist regarding the justification of model based inference for finite populations. Certain authors, for example, Neyman (1971) tend to trust only the man-made randomness imposed in form of the design. Other authors, for example, Barnard (1971), Kalbfleisch and Sprott (1969), Royall (1970a, 1971a), consider inference based on models not only desirable but almost necessary.

In this chapter we treat situations where considerations related to the design, for example, design-unbiasedness, are deemed essential. Typically the model here defines a broad class of distributions. Optimality, under an assumed model, of certain design-unbiased strategies will be established. In Chapters 5 and 6 we treat situations where the design is considered unimportant and the justification for the inferences derives almost entirely from the assumption of a certain model, sometimes narrowly specified as a superpopulation with known shape.

In many situations it is natural to let a model summarize and formalize our prior knowledge about the population, whether this be based on long range experience or on personal subjective belief.

Superpopulation models need not be Bayesian in the sense of expressing personal subjective belief. They can be as objective as some of the models used in classical statistical theory (Royall, 1971b). We thus wish to discourage a tendency in the literature to equate what we shall call a "superpopulation model" with a "Bayesian model."

If $y_1, \ldots, y_N$ were values that have not yet been fixed, it would often be realistic, for a finite population, to consider these numbers as values to be realized of random variables $Y_1, \ldots, Y_N$. We can then describe our uncertainty about what particular values will appear through a probabilistic model, akin to the "objective" models usually used by frequentists.

Now, in the case of a finite population, $y_1, \ldots, y_N$ have usually been fixed already, at some previous point in time. Royall (1971b) argues that it is no less appropriate to use an objective model *after* $y_1, \ldots, y_N$ have been fixed but remain unknown, than *before* they have been fixed.

In the model oriented reasoning of Chapters 5 and 6 we distinguish two main approaches to inference under superpopulation assumptions: (1) models leading to the use of classical (in the sense "non-Bayesian") inference tools; a typical assumption is that $\xi$ contains unknown parameters which must first be estimated, and (2) models leading to the use of Bayesian inference tools; typically the unknown parameters of the model are assigned a prior distribution through the usual methods of assessing priors.

Superpopulation models have a long history in the sampling literature. Early users are Cochran (1939, 1946), Deming and Stephan (1941), Madow and Madow (1944).

As already indicated, the superpopulation concept can be given several interpretations, of which we list a few:

1. The finite population is actually drawn from a larger universe. This is the superpopulation idea in its most pure form.
2. The distribution $\xi$ is modelled to describe a random mechanism or process in the real world; such models are frequently used in econometric or sociometric model building.
3. The distribution $\xi$ is considered as a prior distribution reflecting subjective belief, as in a Bayesian approach. The unobserved numbers among $y_1, \ldots, y_N$ may be looked upon as parameters for which we seek the posterior distribution, given the sample.
4. The distribution $\xi$, while being associated neither with a process in the real world nor with an expression of subjective belief, is used simply as a mathematical device to make explicit the theoretical derivations. For

example, one may be interested in knowing the various model formulations which justify the use of an intuitively appealing estimator such as the sample mean.

5. The Superpopulation approach may be a useful device for incorporating the treatment of nonsampling errors in survey sampling. (This topic is beyond the scope of this book.)

The probabilistic setup in Chapters 4–6 differs in several respects from that of the Fixed population approach of Chapters 2 and 3.

There are two kinds of randomness in the Superpopulation approach. We represent the observed data in the set case as $d = \{(k, y_k) : y \in s\}$. As before, assume that $s$ has been drawn with probability $p(s)$, where the design $p(\cdot)$ is a function on $\mathscr{S}$. (In Chapters 4–6, we consider only the set case. This limitation is justified in view of the sufficiency of $d$ established in Section 2.2.) What is new, in comparison with Chapters 2 and 3, is the introduction of the random variables $Y_1, \ldots, Y_N$ with joint distribution $\xi$ on the Borel sets of $R_N$.

The sample space can be described, as before, as the set of all values of $d$ such that $s \in \mathscr{S}$ and $\mathbf{y} \in R_N$. Note, however, that $\mathbf{y} = (y_1, \ldots, y_N)$ no longer plays the role of a parameter in the sense of Chapters 1–3. Instead, $\xi = \xi_\theta$ may be indexed by a parameter vector $\boldsymbol{\theta}$ belonging to some specified set $\Theta$.

Sections 4.2 and 4.3 give a detailed discussion of new concepts involved under the Superpopulation approach.

## 4.2. A SURVEY OF SUPERPOPULATION MODELS

We now present the most important models to be referred to in the sequel. As mentioned, a "model" defines a *class of distributions* $\xi$. The specification of the class may range from a crude formulation, prescribing, for example only certain features of the means, variances, and covariances of $\xi$, to a situation where $\xi$ is given a highly detailed specification. First we introduce some supplementary notation. Let $\xi$ be an arbitrary superpopulation distribution.

**Definition.** If $Q = Q(Y_1, \ldots, Y_N)$ is a function of $Y_1, \ldots, Y_N$, the $\xi$-*expected value* (or $\xi$-*expectation*) of $Q$, denoted $\mathscr{E}(Q)$, is defined as

$$\mathscr{E}(Q) = \int Q \, d\xi$$

and the $\xi$-*variance* of $Q$, denoted $\mathscr{V}(Q)$, is defined as

$$\mathscr{V}(Q) = \int \{Q - \mathscr{E}(Q)\}^2 \, d\xi$$

If $Q_1 = Q_1(Y_1, \ldots, Y_N)$ and $Q_2 = Q_2(Y_1, \ldots, Y_N)$ are two functions of $Y_1, \ldots, Y_N$, the $\xi$-*covariance* of $Q_1$ and $Q_2$, denoted $\mathscr{C}(Q_1, Q_2)$, is defined as

$$\mathscr{C}(Q_1, Q_2) = \int \{Q_1 - \mathscr{E}(Q_1)\}\{Q_2 - \mathscr{E}(Q_2)\} \, d\xi \qquad \blacksquare$$

In particular, we shall, for $k,l = 1, \ldots, N$, use the notation

$$\mu_k = \mathscr{E}(Y_k)$$
$$\sigma_k^2 = \mathscr{V}(Y_k)$$
$$\sigma_{kl} = \mathscr{C}(Y_k, Y_l) \quad (k \neq l)$$

as well as

$$\bar{\mu} = \frac{\Sigma_1^N \mu_k}{N}$$

We shall distinguish two types of models: Exchangeability models (denoted E) and General models (denoted G). Exchangeability (as defined below) is not required under a G-model. A given model of either type has additional specifications indicated by a subscript of E or of G. Table 4.1 at the end of this section summarizes the essential features of the models to be presented.

***Model $G_T$*** *(transformation model)*. This model defines the class of probability measures $\xi$ on $R_N$ such that, for given numbers $a_k$, $b_k$ $(k = 1, \ldots, N)$, specified by the model maker, the transformed random variables $Z_k = (Y_k - b_k)/a_k$ $(k = 1, \ldots, N)$ have common means, $\mu$, variances, $\sigma^2$, and common covariances, $\rho\sigma^2$, for any pair $k \neq l$. Unless otherwise stated, $\mu$, $\sigma^2$, and $\rho$ are unknown, $-1/(N-1) \leqslant \rho < 1$. Without loss of generality, assume further that the scale factors $a_k$ are positive and normed to satisfy $\Sigma_1^N a_k = N$. (The condition on $\rho$ is needed to ensure that $\mathscr{V}(\bar{Y}) = \sigma^2\{\rho + (1-\rho)A/N\} \geqslant 0$, whatever be the choice of the $a_k$, where $A = \Sigma_1^N a_k^2/N$.)

Under Model $G_T$, the $Y_k$ have the moments:

$$\left.\begin{aligned} \mu_k &= \mathscr{E}(Y_k) = a_k\mu + b_k \\ \sigma_k^2 &= \mathscr{E}(Y_k - \mu_k)^2 = a_k^2\sigma^2 \\ \sigma_{kl} &= \mathscr{E}(Y_k - \mu_k)(Y_l - \mu_l) = a_k a_l \rho\sigma^2 \quad (k \neq l) \end{aligned}\right\} \tag{2.1}$$

Model $G_T$ means that the model-maker is attempting, by choosing the $a_k$ and $b_k$ suitably, to standardize the first two moments of the transformed $Y_k$.

The model assumes that he can assign $a_k$ such that $a_k^2 \propto \mathscr{V}(Y_k)$ for all $k$, and that he can assign $b_k$ such that $\mathscr{E}(Y_k) - b_k \propto \{\mathscr{V}(Y_k)\}^{1/2}$

For example, if the units are farms and $Y$ is acres under wheat, and $b_k = 0$ for $k = 1, \ldots, N$ is assumed, then the large farms would be assigned comparatively speaking large values $a_k$. On the other hand, if all $a_k$ are assumed equal, then the large farms would be assigned large values $b_k$.

In practice, especially if $N$ is large, the sampler would tend to choose the same $a_k, b_k$ for all units belonging to a stratum of units, since the assessment of individual $a_k, b_k$ would be too laborious.

Suppose that the model maker chooses the same number $a_k$ for each unit $k \in V_h$, where $V_h \subset \{1, \ldots, N\}$, meaning that he assesses the same variance, say, $\sigma_h^2$ for each unit in the stratum $V_h$. Moreover, as we see later in this chapter, it makes good theoretical sense to choose a design such that the inclusion probabilities $\alpha_k$ satisfy $\alpha_k \propto a_k$ for $k = 1, \ldots, N$. Thus for any unit $k \in V_h$, the inclusion probability should satisfy $\alpha_k \propto \sigma_h$.

It is then easy to see that a suitable design would be stratified random sampling with stratum sample sizes determined by the well-known optimum allocation rule. For in that design, the number $n_h$ of units to be randomly selected from the $N_h$, say, units in stratum $V_h$ should be such that $n_h \propto N_h \sigma_h$, where $\sigma_h^2$ is the guessed within stratum variance. Noting that for any unit $k \in V_h$ the inclusion probability is $\alpha_k = n_h/N_h$, we see that the condition $\alpha_k \propto \sigma_h$ holds for stratified random sampling with optimal allocation. A more formal discussion of these matters is given in Example 1 of Section 4.4.

Model $G_T$ expresses prior knowledge that is symmetric with respect to the transformed $Y$-variables, in the sense that the first two moments of $Z_1, \ldots, Z_N$ are identical. Among the various special cases of $G_T$, the one where this symmetry applies directly to $Y_1, \ldots, Y_N$ merits a special designation:

***Model*** $G_{T0}$. The special case of Model $G_T$ where $a_k = 1$, $b_k = 0$ for $k = 1, \ldots, N$.

Model $G_{T0}$ expresses the same prior knowledge for all the $Y_k$; in this sense labels are uninformative.

The following regression model is often used in the presence of known auxiliary variable measurements $x_1, \ldots, x_N$.

***Model*** $G_{MR}$ *(multiple regression model).* The class of probability measures $\xi$ on $R_N$ such that $Y_1, \ldots, Y_N$ are independently distributed, and

$$\mu_k = \mathscr{E}(Y_k) = \beta_1 + \sum_{i=2}^{q} \beta_i x_{ki}, \quad \sigma_k^2 = \mathscr{E}(Y_k - \mu_k)^2 = \sigma^2 v_k$$

where $\beta_1, \beta_2, \ldots, \beta_q, \sigma^2$ are unknown, and $x_{k2}, \ldots, x_{kq}, v_k$ is a set of known numbers for every $k$ ($k = 1, \ldots, N$).

The special case of Model $G_{MR}$ where the regression is linear through the origin, on one $x$-variable only, will be considered so often that we denote it separately:

***Model*** $G_R$. The class of probability measures $\xi$ on $R_N$ such that $Y_1, \ldots, Y_N$ are independently distributed, and $\mu_k = \mathscr{E}(Y_k) = \beta x_k$, $\sigma_k^2 = \mathscr{V}(Y_k) = \sigma^2 v(x_k)$

$(k = 1, \ldots, N)$, where $\beta$ and $\sigma^2$ are unknown, $v(\cdot)$ is a known function and $x_1, \ldots, x_N$ are known positive numbers.

A common assumption in Model $G_R$ is that $\sigma_k^2 = \sigma^2 v(x_k) = \sigma^2 x_k^g$, where $g$ is known. In the theoretical discussion of this chapter we treat $g$ as a known constant, but in practice $g$ would most likely be unknown, and the problems caused by this fact are considered in Chapter 7. Cochran (1953, p. 212) and Brewer (1963b) maintain that values of $g$ between 1 and 2 are most frequently borne out by empirical studies. Our assumption will be that the interval of interest is $0 \leqslant g \leqslant 2$.

Note that if in Model $G_T$ we set $a_k^2 \propto v(x_k)$, $b_k = \beta x_k$, $\mu = \rho = 0$, then the moments of Model $G_R$ obtain. But the latter model is not in general a special case of the former, which assumes the $b_k$ to be known. In Model $G_R$, $\beta$, and hence the $\beta x_k$, are unknown. (In the case of $v(x) = x^2$, Model $G_R$ does satisfy the conditions of Model $G_T$, namely, if we let $a_k = x_k/\bar{x}$, $b_k = 0$ and $\rho = 0$.)

In the models considered so far, no assumptions were made about the shape of the (marginal) distributions of $\xi$. If such assumptions can legitimately be made, the models can, of course, be expanded to include this information.

Next, we consider various types of *exchangeability models*. In order to be exchangeable, the distribution $\xi$ must be symmetric in accordance with the following definition.

**Definition.** Random variables $Y_1, \ldots, Y_N$ are called *exchangeable* if $Y_{r_1}, \ldots, Y_{r_N}$ has, for every permutation $r_1, \ldots, r_N$ of $1, \ldots, N$, the same joint distribution, which is called an *exchangeable distribution*. ■

The concept of exchangeability was introduced by de Finetti (1937); a discussion of symmetric distributions is found in Hewitt and Savage (1955). In inference from finite populations, the notion was popularized in the Bayesian framework by Ericson (1965, 1969a,b) under the name of "exchangeable prior." The usefulness of exchangeability is, of course, not limited to the Bayesian mode of inference.

Let $\pi = (\pi(1), \ldots, \pi(N))$ denote a permutation of $1, \ldots, N$, and let $F_\pi(\mathbf{y})$ and $F(\mathbf{y})$ denote the joint distribution functions of $Y_{\pi(1)}, \ldots, Y_{\pi(N)}$ and of $Y_1, \ldots, Y_N$, respectively.

If $\pi^{-1} = (r_1, \ldots, r_N)$ denotes the permutation such that $\pi(r_k) = k$ $(k = 1, \ldots, N)$, then the two events $Y_{\pi(k)} \leqslant y_k$ $(k = 1, \ldots, N)$ and $Y_k \leqslant y_{r_k}$ $(k = 1, \ldots, N)$ are identical. Hence, if $F(\cdot)$ is exchangeable,

$$F(\mathbf{y}) = F_\pi(\mathbf{y}) = F(\mathbf{y}_{\pi^{-1}})$$

where $\mathbf{y}_{\pi^{-1}} = (y_{r_1}, \ldots, y_{r_N})$. That is, $F(\cdot)$ is a symmetric function in its $N$ arguments.

As pointed out by Ericson (1969a), exchangeability expresses the prior knowledge that the labels, although observable, carry no information about the value $y_k$ associated with label $k$. He feels that the idea of an exchangeable distribution $\xi$ for $Y_1, \ldots, Y_N$ approximates what many practitioners have in mind when they consider simple random sampling to be appropriate.

The variables $Y_1, \ldots, Y_N$ *themselves* may be assumed to be exchangeable. Frequently, however, it is convenient to assume that, following a location and scale transformation, the *transformed* $Y_k$ are exchangeable.

The following Model $E_T$ is of this kind. The first and second moments are as in Model $G_T$, but exchangeability is an added requirement. In other words, Model $E_T$ implies Model $G_T$, but not vice versa.

***Model $E_T$.*** This model defines the class of probability measures $\xi$ on $R_N$ such that, for known numbers $a_k > 0, b_k$ $(k = 1, \ldots, N)$ satisfying

$$\Sigma_1^N a_k = N$$

the random variables $Z_k = (Y_k - b_k)/a_k$ $(k = 1, \ldots, N)$ have an exchangeable, absolutely continuous distribution. The common means, variances and covariances implied by the exchangeability will be denoted by $\mu$, $\sigma^2$, and $\rho\sigma^2$, respectively. The first and second moments of the $Y_k$ are then given by (2.1).

The $Y_k$ themselves become exchangeable in an important special case:

***Model $E_{T0}$.*** The special case of Model $E_T$ such that $a_k = 1$, $b_k = 0$ $(k = 1, \ldots, N)$.

Consider next the discrete exchangeable superpopulations known as random labeling (or random permutation) models. This idea, which found an early application in Madow and Madow (1944), was shown more recently by Kempthorne (1969) to be useful in discussing inference from finite populations.

In its simplest form, the model can be described as the assumption that the units, which carry the fixed but unknown numbers $y_1, \ldots, y_N$, have been labeled at random. Each permutation $\pi = (\pi(1), \ldots, \pi(N))$ of $1, \ldots, N$ is assumed to have the same probability, $1/N!$, of being assigned as labels for the units. Equivalently, one could say that the fixed but unknown numbers $y_1, \ldots, y_N$ have been randomly assigned to the units, which are labeled from 1 to $N$, each permutation of $y_1, \ldots, y_N$ having probability $1/N!$ of being assigned to the fixed labels $1, \ldots, N$. Hence there is no systematic relationship between the labels and the associated $y_k$-values.

The model can be used in a more general form: Let there be $N$ fixed but unknown numbers $z_1, \ldots, z_N$. Assign the numbers $z_1, \ldots, z_N$ at random to the units, which wear the fixed labels $1, \ldots, N$, in such a way that each permutation of $z_1, \ldots, z_N$ has probability $1/N!$ of being assigned to the fixed labels. Then no systematic relationship exists between the labels and the $z_k$-values. The value that ends up assigned to label $k$ can be seen as the outcome of a random variable, say, $Z_k$. Moreover, assume that $Z_k = (Y_k - b_k)/a_k$, where

$a_k > 0, b_k$ are numbers tied to label $k$ as in Model $G_T$, and $\Sigma_1^N a_k = N$. The random assignment assumption implies a certain distribution for $Y_1, \ldots, Y_N$.

**Example 1.** Let $N = 3$, $a_1 = 3/4$, $a_2 = 1$, $a_3 = 5/4$, and $b_1 = b_2 = b_3 = 0$. The random variables are $Z_1 = 4Y_1/3$, $Z_2 = Y_2$, and $Z_3 = 4Y_3/5$. Under the random permutation model, each of the six possible permutations of $z_1, z_2, z_3$ are equally likely outcomes for $(Z_1, Z_2, Z_3)$. For example, for the permutation $z_3, z_2, z_1$ we have

$$P(Z_1 = z_3, Z_2 = z_2, Z_3 = z_1) = 1/6,$$

or

$$P(Y_1 = 3z_3/4, Y_2 = z_2, Y_3 = 5z_1/4) = 1/6.$$

In other words, one of the six equally likely populations under this model is described by the vector $\mathbf{y} = (3z_3/4, z_2, 5z_1/4)$. To the permutation $z_1, z_2, z_3$ corresponds $\mathbf{y} = (3z_1/4, z_2, 5z_3/4)$, and so on. ■

This description fits the following model formulation:

*Model $E_{RP}$ (random permutation model).* The class of distributions $\xi$ such that, for any fixed, unknown numbers $z_1, \ldots, z_N$, and for given numbers $a_k > 0$, $b_k$ $(k = 1, \ldots, N)$ such that $\Sigma_1^N a_k = N$, the random variables $Z_k = (Y_k - b_k)/a_k$ have an exchangeable distribution such that

$$P(Z_1 = z_{r_1}, \ldots, Z_N = z_{r_N}) = 1/N!$$

for each permutation $r_1, \ldots, r_N$ of $1, \ldots, N$.

Model $E_{RP}$, in which the fixed vector $(z_1, \ldots, z_N)$ acts as a parameter indexing $\xi$, represents an expression of the prior belief that "the labels are uninformative" with respect to the $z_k$. The model has been used, for example, with $a_k \propto x_k^r, b_k = 0$ $(k = 1, \ldots, N)$, where $x_k$ is the known value of an auxiliary variable, and $r$ is a constant; see C. R. Rao (1971), Godambe and Thompson (1973), J. N. K. Rao (1975).

Model $E_{RP}$ implies, for any $n$ such that $1 \leqslant n \leqslant N$, the marginal distribution

$$P(Z_{k_1} = z_{r_1}, \ldots, Z_{k_n} = z_{r_n}) = 1/N^{(n)}$$

for each of $N^{(n)} = n! \binom{N}{n}$ different sequences $r_1, \ldots, r_n$ of $n$ numbers chosen from $z_1, \ldots, z_N$, where $Z_{k_1}, \ldots, Z_{k_n}$ are the random variables associated with an arbitrary, fixed subset of labels $k_1, \ldots, k_n$.

The $\xi$-moments of the $Z_k$ are, for $k = 1, \ldots, N$:

$$\mathcal{E}(Z_k) = \mu_z$$

$$\mathcal{E}(Z_k - \mu_z)^2 = \sigma_z^2$$

$$\mathcal{E}(Z_k - \mu_z)(Z_l - \mu_z) = \frac{-\sigma_z^2}{N-1} \qquad (k \neq l)$$

where the unknown $\mu_z$ and $\sigma_z^2$ are given by

$$\mu_z = \frac{\Sigma_1^N z_k}{N}, \quad \sigma_z^2 = \frac{\Sigma_1^N (z_k - \mu_z)^2}{N} \tag{2.2}$$

The implied moments for the $Y_k$, which are generally not exchangeable, are of the form (2.1), namely,

$$\mu_k = \mathscr{E}(Y_k) = a_k \mu_z + b_k$$

$$\sigma_k^2 = \mathscr{E}(Y_k - \mu_k)^2 = a_k^2 \sigma_z^2$$

$$\sigma_{kl} = \mathscr{E}(Y_k - \mu_k)(Y_l - \mu_l) = \frac{-a_k a_l \sigma_z^2}{N-1}$$

We list separately the special case of Model $E_{RP}$ where the $Y_k$ themselves are exchangeable. This occurs when $a_k = 1, b_k = 0$ $(k = 1, \ldots, N)$, so that $Z_k = Y_k$ in Model $E_{RP}$. We assume that the $N!$ permutations of fixed, unknown numbers $y_1, \ldots, y_N$ constitute equally likely outcomes for the vector random variable $(Y_1, \ldots, Y_N)$.

***Model*** $E_{RP0}$. The special case of Model $E_{RP}$ such that $a_k = 1, b_k = 0$ $(k = 1, \ldots, N)$, $\mu_z = \mu_y = \Sigma_1^N y_k/N$, $\sigma_z^2 = \sigma_y^2 = \Sigma_1^N (y_k - \mu_y)^2/N$.

Finally, we consider some superpopulation models $\xi$ which may be termed "parametric": Usually they assume a known shape of the joint distribution of $Y_1, \ldots, Y_N$, which is indexed by a (usually unknown) parameter vector $\boldsymbol{\theta} = (\theta_1, \ldots, \theta_r)$.

In the case of absolutely continuous distributions, let $g(\mathbf{y} \mid \boldsymbol{\theta})$ denote the density function of $Y_1, \ldots, Y_N$, where $\mathbf{y} = (y_1, \ldots, y_N)$ and $\boldsymbol{\theta} \in \Theta$, the parameter space.

We list three cases of particular interest in the sequel; the subscripts $P$ and $PI$ may be interpreted as "parametric" and "parametric independent," respectively:

***Model*** $G_{PI}$. The class of absolutely continuous distributions $\xi$ such that $Y_1, \ldots, Y_N$ are independently but not necessarily identically distributed, their joint density function being

$$g(\mathbf{y} \mid \boldsymbol{\theta}) = \Pi_1^N g_k(y_k \mid \boldsymbol{\theta}), \quad \boldsymbol{\theta} \in \Theta$$

where $g_k(\cdot \mid \boldsymbol{\theta})$ is the density function of $Y_k$.

***Model*** $E_P$. The class of absolutely continous distributions $\xi$ such that $Y_1, \ldots, Y_N$ are exchangeable, their joint density function, symmetric in its $N$ arguments, being

$$g(\mathbf{y} \mid \boldsymbol{\theta}), \quad \boldsymbol{\theta} \in \Theta$$

*Model* $E_{PI}$. The class of absolutely continuous distributions $\xi$ such that $Y_1, \ldots, Y_N$ are independently and identically distributed (and hence exchangeable), their joint density function being

$$g(\mathbf{y} \mid \boldsymbol{\theta}) = \Pi_1^N g(y_k \mid \boldsymbol{\theta}), \quad \boldsymbol{\theta} \in \Theta \tag{2.3}$$

where $g(\cdot \mid \boldsymbol{\theta})$ is the density function common to all the $Y_k$.

We note that Models $E_P$ and $E_{PI}$ can be considered as more detailed specifications of Model $E_{T0}$. Similarly, Model $G_{PI}$ may be considered an elaboration of Model $G_{T0}$.

**Remark 1.** In the prediction approach based on classical inference tools, Chapter 5, the inference problem at hand may be described as prediction of $\bar{Y}$, a random variable with a distribution determined by $\xi$. This turns out to involve estimation, in the classical sense of this term, of (some function of) the unknown $\boldsymbol{\theta}$ in $g(\mathbf{y} \mid \boldsymbol{\theta})$. ■

Finally, we observe that a new superpopulation distribution (referred to by Bayesians as a "prior distribution for $y_1, \ldots, y_N$") can be generated as a mixing distribution, namely, if we assess a prior distribution for $\boldsymbol{\theta}$, $F(\boldsymbol{\theta} \mid \boldsymbol{\phi})$, where $\boldsymbol{\phi} = (\phi_1, \ldots, \phi_t)$ is a known parameter vector. Letting, as before, $g(\cdot \mid \boldsymbol{\theta})$ be the density of the $Y_k$ conditional on $\boldsymbol{\theta}$ we arrive at

$$h(\mathbf{y} \mid \boldsymbol{\phi}) = \int_{\Theta} g(\mathbf{y} \mid \boldsymbol{\theta}) \, dF(\boldsymbol{\theta} \mid \boldsymbol{\phi}) \tag{2.4}$$

If $g(\cdot \mid \boldsymbol{\theta})$ is exchangeable, so is the resulting $h(\cdot \mid \boldsymbol{\phi})$. In particular, an exchangeable distribution is generated from (2.4) if, as did Aggarwal (1959, 1966) and Ericson (1965, 1969a,b), we take the $Y_k$ to be independently and identically distributed, conditional on $\boldsymbol{\theta}$, that is, $g(\mathbf{y} \mid \boldsymbol{\theta})$ in (2.4) is of the form (2.3).

**Remark 2.** In the Bayesian inference situations of Chapter 6, $h(\cdot \mid \boldsymbol{\phi})$ plays the role of prior distribution for the unknown vector $\mathbf{y} = (y_1, \ldots, y_N)$. Via the posterior of $y_1, \ldots, y_N$, given the sample data, we can make Bayesian inference about the unknown quantity $\bar{y} = \Sigma_1^N y_k/N$. A somewhat different interpretation of this situation is to consider that $Y_1, \ldots, Y_N$ have the superpopulation distribution $g(\mathbf{y} \mid \boldsymbol{\theta})$, conditional on the unknown $\boldsymbol{\theta}$, which is assigned a completely specified prior $F(\boldsymbol{\theta} \mid \quad)$. The same inference problem may then be described as prediction of the random variable $\bar{Y}$, which will turn out to involve Bayesian estimation of (some function of) the unknown $\boldsymbol{\theta}$. ■

For ready future reference, Table 4.1 summarizes the models introduced so far. Recall that the notation is as follows: $\mu_k = \mathscr{E}(Y_k)$, $\sigma_k^2 = \mathscr{V}(Y_k) = \mathscr{E}(Y_k - \mu_k)^2$, $\sigma_{kl} = \mathscr{C}(Y_k, Y_l) = \mathscr{E}(Y_k - \mu_k)(Y_l - \mu_l)$ $(k \neq l)$. For given $a_k, b_k$

**Table 4.1. A Summary of Superpopulation Models**

| Model | $\mu_k$ | $\sigma_k^2$ | $\sigma_{kl}$ | Remarks |
|---|---|---|---|---|
| $G_T$ | $a_k\mu + b_k$ | $a_k^2\sigma^2$ | $a_k a_l \rho\sigma^2$ | Transformation model: The $Z_k = (Y_k - b_k)/a_k$ have common unknown $\mu$, $\sigma$, $\rho$. |
| $G_{T0}$ | $\mu$ | $\sigma^2$ | $\rho\sigma^2$ | The $Y_k$ have common, unknown $\mu$, $\sigma$, $\rho$. |
| $G_{MR}$ | $\beta_1 + \Sigma_2^q \beta_i x_{ki}$ | $\sigma^2 v_k$ | 0 | Multiple regression model: The $Y_k$ are independent; $\beta_1, \ldots, \beta_q$ and $\sigma$ unknown. |
| $G_R$ | $\beta x_k$ | $\sigma^2 v(x_k)$ | 0 | Regression model: The $Y_k$ are independent; $\beta$, $\sigma$ unknown. |
| $E_T$ | $a_k\mu + b_k$ | $a_k^2\sigma^2$ | $a_k a_l \rho\sigma^2$ | Exchangeable transformation model: The $Z_k$ are exchangeable; $\mu$, $\sigma$, $\rho$ unknown. |
| $E_{T0}$ | $\mu$ | $\sigma^2$ | $\rho\sigma^2$ | The $Y_k$ are exchangeable; $\mu$, $\sigma$, $\rho$ unknown. |
| $E_{RP}$ | $a_k\mu_z + b_k$ | $a_k^2\sigma_z^2$ | $\dfrac{-a_k a_l \sigma_z^2}{N-1}$ | Random permutation model: The $Z_k$ are exchangeable; $\mu_z$, $\sigma_z$ unknown. |
| $E_{RP0}$ | $\mu_y$ | $\sigma_y^2$ | $\dfrac{-\sigma_y^2}{N-1}$ | The $Y_k$ are exchangeable; $\mu_y$, $\sigma_y$ unknown. |
| $G_{PI}$ | $\mu_k(\boldsymbol{\theta})$ | $\sigma_k^2(\boldsymbol{\theta})$ | 0 | Parametric model: The $Y_k$ are independent; $\boldsymbol{\theta}$ usually unknown. |
| $E_P$ | $\mu(\boldsymbol{\theta})$ | $\sigma^2(\boldsymbol{\theta})$ | $\rho(\boldsymbol{\theta})\sigma^2(\boldsymbol{\theta})$ | Parametric model: The $Y_k$ are exchangeable; $\boldsymbol{\theta}$ usually unknown. |
| $E_{PI}$ | $\mu(\boldsymbol{\theta})$ | $\sigma^2(\boldsymbol{\theta})$ | 0 | Parametric model: The $Y_k$ are independent and identically distributed; $\boldsymbol{\theta}$ usually unknown. |

such that $a_k > 0$ $(k = 1, \ldots, N)$ and $\Sigma_1^N a_k = N$, $Z_k = (Y_k - b_k)/a_k$ denote transformed $Y$-variables.

## 4.3. SOME TERMINOLOGY

The objective is to estimate $\bar{y} = \Sigma_1^N y_k/N$, the population mean. Now, $\mathbf{y} = (y_1, \ldots, y_N)$ is regarded as an outcome of $\mathbf{Y} = (Y_1, \ldots, Y_N)$ with distribution $\xi$, and $\bar{y}$ is an outcome of $\bar{Y} = \Sigma_1^N Y_k/N$.

When superpopulation models are involved, the available data are $d = \{(k, y_k)\colon k \in s\}$, where $s \in \mathscr{S}$ and $y_k \in R_1$ for $k = 1, \ldots, N$. The data $d$ is the observed outcome of the random variable $\mathscr{D} = \{(k, Y_k) : k \in S\}$, where $S$ is random, and also, for each value $s$ of $S$, $Y_k$ for $k \in s$ is random.

Two additional random variables are recognized: $D = \{(k, y_k) : k \in S\}$, which, as used in Chapters 2 and 3, is the random variable obtained from $\mathscr{D}$ if $Y_k$ is fixed at $y_k$ for $k = 1, \ldots, N$, and also $\mathscr{d} = \{(k, Y_k) : k \in s\}$, which is the random variable obtained from $\mathscr{D}$ if $S$ is fixed at $s$.

The *sample space* of $\mathscr{D}$ taking values $d$ is

$$\mathscr{X} = \{d : s \in \mathscr{S}, \mathbf{y} \in \Omega\}$$

where usually $\Omega = R_N$. This sample space is thus identical to (5.4) of Section 1.5.

Consider an outcome $S = s = \{k_1, \ldots, k_{\nu(s)}\}$ where $k_1 < \ldots < k_{\nu(s)}$ is an enumeration of the labels in $s$. Let $g_s(y_{k_1}, \ldots, y_{k_{\nu(s)}}; \boldsymbol{\theta})$ be the marginal density of $Y_{k_1}, \ldots, Y_{k_{\nu(s)}}$

If the design is noninformative, as defined in Section 1.4, then the probability density of the outcome $\mathscr{D} = d$ is

$$p_{\mathscr{D}}(d) = p(s) g_s(y_{k_1}, \ldots, y_{k_{\nu(s)}}; \boldsymbol{\theta}) \tag{3.1}$$

A *statistic* is a function $T = T(\mathscr{D})$ such that, for any given value $s$ of $S$, $T$ depends on $Y_1, \ldots, Y_N$ only through those $Y_k$ for which $k \in s$.

When the statistic $T(\mathscr{D})$ is used for making inference about the population mean, we shall call $T(\mathscr{D})$ *a predictor* of the random variable $\bar{Y}$. The random variable $T(\mathscr{d})$ obtained from $T(\mathscr{D})$ for $S = s$ will also be referred to as a *predictor* of $\bar{Y}$: It is still a function of the random variables $Y_k : k \in s$. To indicate a predictor, we shall often write simply $T$ in place of either $T(\mathscr{D})$ or $T(\mathscr{d})$, and it should be kept in mind that $T$, the capital letter, is always a function of the random variables $Y_k$, where $k \in S$ or $k \in s$.

The random variable obtained from $T(\mathscr{D})$ if $Y_k$ is fixed at $y_k$ for $k \in S$ will be written $t(D)$. The value of $T(\mathscr{D})$ for the outcome $S = s$ and $Y_k = y_k$ for $k \in s$ will be written $t(d)$; this is the "estimate" or "predicted value" of $\bar{y}$. Often, however, we shall simply write $t$ in place of either $t(D)$ or $t(d)$, the small letter $t$ indicating that $t$ is a function of the realized values $y_k$ ot the $Y_k$ for $k \in S$ or $k \in s$.

The operators $E(\cdot)$, $V(\cdot)$, and $\text{MSE}(\cdot)$ used in earlier chapters will be retained to denote $p$-expectation, $p$-variance, and $p$-MSE, respectively. Thus, if $T$ is a predictor, both

$$E(T) = \Sigma_{\mathscr{S}}\, p(s) T$$

and

$$\text{MSE}(T) = \Sigma_{\mathscr{S}}\, p(s) (T - \bar{Y})^2$$

are functions of the random variables $Y_1, \ldots, Y_N$.

The following definitions relate to properties of unbiasedness of a predictor $T$:

**Definition.** $T$ is called a *p-unbiased* predictor of $\bar{Y}$ if and only if, for a given design $p, E(t) = \bar{y}$ for all $\mathbf{y} = (y_1, \ldots, y_N) \in R_N$, where $t$ is the value of $T$ for $Y_k = y_k$, $k \in S$. The strategy $(p, T)$ is called $p$-unbiased if $T$ is a $p$-unbiased predictor under $p$.

$T$ is called a *ξ-unbiased* predictor of $\bar{Y}$ if and only if, for a given $\xi$, $\mathscr{E}(T - \bar{Y}) = 0$ for all $s \in \mathscr{S}$.

$T$ is called a *pξ-unbiased* predictor of $\bar{Y}$ if and only if, for given $p$ and $\xi$, $\mathscr{E}E(T - \bar{Y}) = 0$. ■

**Remark 1.** FOr convenience we shall sometimes call $T$ a $\xi$-unbiased *estimator* of $\mathscr{E}(\bar{Y}) = \bar{\mu} = \Sigma_1^N \mu_k/N$, namely if and only if $T$ is a $\xi$-unbiased predictor of $\bar{Y}$. Similarly, we shall sometimes call $T$ a $p\xi$-unbiased *estimator* of $\bar{\mu}$ if and only if $T$ is a $p\xi$-unbiased predictor of $\bar{Y}$. Also, the shortcut expressions "$T$ is $p$-unbiased ($\xi$-unbiased, $p\xi$-unbiased)" should be understood in the light of the above definition and the present remark. ■

**Remark 2.** For any given design $p$ and given model $\xi$, the class of $p\xi$-unbiased predictors contains the class of $p$-unbiased predictors as well as the class of $\xi$-unbiased predictors. ■

The topic of the present chapter is $p$-unbiased estimation of $\bar{y}$, seen in the light of a superpopulation model. In Chapters 5 and 6 the design aspect is de-emphasized and the use of a given predictor will be justified by the assumed validity of a certain superpopulation model; $\xi$-unbiased predictors will, for example, be a topic of primary interest.

In Chapters 4–6, the choice of a strategy $(p, T)$ will be dictated by the objective to minimize the $\xi$-expected $p$-MSE,

$$\mathscr{E}\text{MSE}(p, T) = \mathscr{E}E(T - \bar{Y})^2$$

which reduces, if $T$ is $p$-unbiased, to the $\xi$-expected $p$-variance, $\mathscr{E}V(p, T)$. The notation $\mathscr{E}\text{MSE}(p, T)$ and $\mathscr{E}V(p, T)$ emphasizes that the MSE or $V$ operator is with respect to the specific design $p$.

Among various possible criteria for judging the uncertainty of strategies, $\mathscr{E}\text{MSE}(p, T) = \mathscr{E}E(T - \bar{Y})^2$ is the most fully explored one. Since $\bar{Y}$ is the character of principal interest, it is reasonable that the criterion measure average squared deviation from $\bar{Y}$. However, if our interest were in making inference *directly* to the superpopulation, averaging of $(T - \bar{\mu})^2$ would be a more natural criterion. In some practical situations, $\bar{\mu}$ (rather than $\bar{Y}$) may indeed bε the natural target of inference. Lemmas 4.2 and 4.5, and the predictor $T^\circ$ given by (2.5) of Section 5.2, refer explicitly to inference about $\bar{\mu}$, but beyond this our principal goal will be to predict $\bar{Y}$.

Although $\mathscr{E}E(T-\bar{Y})^2$ will be the criterion of prime interest in Chapters 4–6, the outlook is different in each chapter.

In this chapter, design oriented aspects dominate, which would motivate consideration of $E(T-\bar{Y})^2$. The additional averaging with respect to a superpopulation $\xi$ is a device often used in classical survey sampling (see Cochran, 1953, p. 169; 1963, p. 214) to resolve a mathematical difficulty: Comparison of variances of two $p$-unbiased strategies is often difficult for any one finite population, but easy on the average over a series of finite populations generated by a given distribution $\xi$. Minimization of $\mathscr{E}E(T-\bar{Y})^2$ under $p$-unbiasedness means that we require $T$ to be $p$-unbiased for every one of the populations that can be generated by $\xi$. We shall find in this chapter that consideration of $\mathscr{E}E(T-\bar{Y})^2$ helps to resolve some of the nonexistence problems in UMV $p$-unbiased estimation encountered in Chapter 3.

Adherents of the model-based approach argue that, since a model is invoked in the first place, $p$-unbiasedness is an unnecessarily heavy restriction; instead, $\xi$-unbiasedness or possibly $p\xi$-unbiasedness should be required. They would also argue that averaging of $(T-\bar{Y})^2$ with respect to the design $p$ is a matter of presampling interest only. For further discussion of these matters, see Section 5.5, and Cassel, Särndal and Wretman (1976), Smith (1976).

Chapters 5–6 deal with the model-based approach, that is, the approach where the $\xi$ distribution is the essential element for inference, where $s$ is treated as given, and where the design $p$ that produced $s$ is of minor interest. The first objective in Chapters 5–6 is therefore to choose $T$, for every given $s$, to minimize $\mathscr{E}(T-\bar{Y})^2$. Beyond this, averaging with respect to $p$ is of secondary importance. (However, it turns out that the predictor $T$ that minimizes $\mathscr{E}(T-\bar{Y})^2$ for any given $s$ is also the predictor that minimizes $\mathscr{E}\mathrm{E}(T-\bar{Y})^2$ for any given *noninformative* design $p$; see Section 5.1.) Thus in the model-based approach, the problem of choice of a good strategy $(p, T)$ is less important than the problem of choice of a predictor $T$ which is good for any sample $s$ actually obtained.

**Remark 3.** If $T_1$ and $T_2$ are predictors such that, for the given design $p$, $\mathscr{E}\mathrm{MSE}(p, T_1) \leqslant \mathscr{E}\mathrm{MSE}(p, T_2)$ for all $\xi \in \mathscr{C}$, a given class of superpopulations, then we shall say that $T_1$ is *at least as good* a predictor as $T_2$ for the design $p$. If strict inequality holds for at least one $\xi \in \mathscr{C}$, then $T_1$ will be called *better* than $T_2$.

If $(p_1, T_1)$ and $(p_2, T_2)$ are strategies such that $\mathscr{E}\mathrm{MSE}(p_1, T_1) \leqslant \mathscr{E}\mathrm{MSE}(p_2, T_2)$ for all $\xi \in \mathscr{C}$ then we shall say that $(p_1, T_1)$ is *at least as good* a strategy as $(p_2, T_2)$. If strict equality holds for at least one $\xi \in \mathscr{C}$, then we shall say that $(p_1, T_1)$ is *better* than $(p_2, T_2)$. ■

**Remark 4.** If $t_1 = t_1(D)$ and $t_2 = t_2(D)$ are estimators of $\bar{y}$ in the sense of Chapter 1, and $E(t_1-\bar{y})^2 \leqslant E(t_2-\bar{y})^2$ for all $\mathbf{y} \in R_N$, then $\mathscr{E}E(T_1-\bar{Y})^2$

$\leqslant \mathscr{E}E(T_2 - \bar{Y})^2$ for any superpopulation $\xi$, where $T_1$ and $T_2$ are the predictors corresponding, respectively, to the estimators, $t_1$ and $t_2$. Hence if $t_1$ is at least as good as $t_2$ as an estimator of $\bar{y}$, for a given design $p$, then $T_1$ is at least as good a predictor as $T_2$, for any $\xi$. ■

The following lemma gives a useful partition of $\mathscr{E}\,\text{MSE}(p, T)$.

**Lemma 4.1.** *Let $T$ be any predictor of $\bar{Y}$. For any $\xi$, and for any noninformative $p$.*

$$\mathscr{E}\text{MSE}(p, T) = E\mathscr{V}(T) + E\{\mathscr{B}(T)\}^2 + \mathscr{V}(\bar{Y}) - 2\mathscr{E}\{(\bar{Y} - \bar{\mu})E(T - \bar{\mu})\} \quad (3.2)$$

*where $\mathscr{V}(T) = \mathscr{E}\{T - \mathscr{E}(T)\}^2$ and $\mathscr{B}(T) = \mathscr{E}(T - \bar{Y})$ are the $\xi$-variance and the $\xi$-bias, respectively, of $T$. In particular*

*(a) If $T$ is $p$-unbiased:*

$$\mathscr{E}V(T) = E\mathscr{V}(T) + E\{\mathscr{B}(T)\}^2 - \mathscr{V}(\bar{Y}) \quad (3.3)$$

*(b) If $T$ is $p$- as well as $\xi$-unbiased:*

$$\mathscr{E}V(T) = E\mathscr{V}(T) - \mathscr{V}(\bar{Y}) \quad (3.4)$$

**Proof.** As $p$ is noninformative, (see Section 1.4), the order between the operators $E$ and $\mathscr{E}$ may be changed. The equalities

$$\mathscr{E}E(T - \bar{Y})^2 = E\mathscr{E}(T - \bar{\mu})^2 + \mathscr{V}(\bar{Y}) - 2\mathscr{E}\{(\bar{Y} - \bar{\mu})E(T - \bar{\mu})\}$$

and $\mathscr{E}(T - \bar{\mu})^2 = \mathscr{V}(T) + \{\mathscr{B}(T)\}^2$ produce the desired result. The special cases (a) and (b) follow by noting that $E(T) = \bar{Y}$ under $p$-unbiasedness and that $\mathscr{B}(T) = 0$ under $\xi$-unbiasedness. □

**Remark 5.** Formula (3.2) may also be written as

$$\mathscr{E}\text{MSE}(p, T) = E\mathscr{V}(T) + E\{\mathscr{B}(T)\}^2 + \mathscr{E}\{B(T)\}^2 - \mathscr{E}\{E(T) - \bar{\mu}\}^2$$

where $B(T) = E(T) - \bar{Y}$ is the $p$-bias of $T$. The expression follows by noting that

$$\mathscr{E}\{B(T)\}^2 = \mathscr{E}\{E(T) - \bar{\mu}\}^2 + \mathscr{V}(\bar{Y}) - 2\mathscr{E}\{(\bar{Y} - \bar{\mu})E(T - \bar{\mu})\} \quad ■$$

## 4.4. OPTIMAL DESIGN-UNBIASED STRATEGIES: MODEL $G_T$

In the remainder of this chapter, assume that the predictors $T$ under consideration are $p$-unbiased. The problem in this chapter is to find the optimal $p$-unbiased strategy $(p, T)$, that is, joint optimization with respect to $T$ and $p$ is involved. Consider first Model $G_T$. Theorem 4.1 below shows that the $p$-unbiased strategy that uses the generalized difference predictor is optimal among strategies based on linear (homogeneous or nonhomogeneous) predictors. Theorem 4.2 of the following section removes the linearity restriction, but requires stronger conditions on the model.

Given an arbitrary vector $\mathbf{e} = (e_1, \ldots, e_N)$, and a design with inclusion probabilities $\alpha_k > 0$ $(k = 1, \ldots, N)$, the generalized difference predictor is given by

$$T_{GD} = \Sigma_S \frac{Y_k - e_k}{N\alpha_k} + \bar{e} \tag{4.1}$$

where $\bar{e} = \Sigma_1^N e_k/N$. We recall the following properties of $T_{GD}$:

(i) $T_{GD}$ is $p$-unbiased;
(ii) $T_{GD}$ has zero $p$-variance for any value y of $\mathbf{Y}$ that satisfies $(\mathbf{y} - \mathbf{e}) \propto \boldsymbol{\alpha} = (\alpha_1, \ldots, \alpha_N)$, provided that $p$ is an FES($n$) design;
(iii) $T_{GD}$ is $\xi$-unbiased for any model if $e_k = \mu_k$ $(k = 1, \ldots, N)$;
(iv) if

$$T_{HT} = \Sigma_S \frac{Y_k}{N\alpha_k} \tag{4.2}$$

denotes the Horvitz–Thompson predictor, and if $e_k = c\alpha_k$ $(k = 1, \ldots, N)$, for an arbitrary constant $c$, then

$$T_{GD} = T_{HT} - \frac{c[\nu(s) - E\{\nu(S)\}]}{N}$$

Thus, $T_{GD}$ reduces to $T_{HT}$ (i) if $\mathbf{e} = \mathbf{0}$ or (ii) if $e_k \propto \alpha_k$ $(k = 1, \ldots, N)$ and $p$ is an FES design;
(v) $T_{GD}$ is origin and scale invariant; see Section 3.7.

If $p$ is an FES($n$) design with $\alpha_k > 0$ $(k = 1, \ldots, N)$, then $\mathscr{E} V(p, T_{GD})$ is minimized for the choice $e_k = \mu_k$ and $\alpha_k = fa_k$ $(k = 1, \ldots, N)$, where $f = n/N$. Hence, the optimal strategy of type $(p, T_{GD})$ is given by $(p_o, T_{GDo})$, consisting of

(i) any FES($n$) design

$$p_o = p_o(s) \text{ such that } \alpha_k = fa_k \quad (k = 1, \ldots, N) \tag{4.3}$$

(ii) the predictor

$$T_{GDo} = \Sigma_S \frac{Y_k - b_k}{na_k} + \bar{b} \tag{4.4}$$

where $\bar{b} = \Sigma_1^N b_k/N$.

Contained in Theorem 4.1 below, this result requires no separate proof. The predictor $T_{GDo}$ has the following properties:

(i) $T_{GDo}$ is $\xi$-unbiased and $p\xi$-unbiased for any $\xi$ satisfying Model $G_T$ and for any FES($n$) design $p$,

(ii) $T_{GDo}$ is $p$-unbiased under $p = p_o$, but for an arbitrary FES($n$) design $T_{GDo}$ is not necessarily $p$-unbiased.

Consider predictors $T$ such that $T \in \mathscr{L}_u$, the class of all $p$-unbiased linear predictors of $\bar{Y}$. Hence, $E(T) = \bar{Y}$ and $T$ is of the form

$$T = w_{0s} + \Sigma_s w_{ks} Y_k \tag{4.5}$$

Theorem 4.1 due to Cassel, Särndal, and Wretman (1976), states that, under Model $G_T$, $(p_o, T_{GDo})$ is the optimal $p$-unbiased strategy among all $(p, T)$ such that $T \in \mathscr{L}_u$ and $p$ is any FES design. The proof is simplified if we consider first the following lemma dealing with $p\xi$-unbiased estimation of $\bar{\mu}$ for any given FES design $p$.

**Lemma 4.2.** *Let p be any given FES(n) design with $\alpha_k > 0$ $(k = 1, \ldots, N)$. Then, under Model $G_T$,*

$$\mathscr{E}E(T - \bar{\mu})^2 \geqslant \mathscr{E}E(T_{GDo} - \bar{\mu})^2 = \frac{\{1 + \rho(n-1)\}\sigma^2}{n}$$

*for any linear $p\xi$-unbiased estimator $T$ of $\bar{\mu}$, that is, $T$ is of the form (4.5) and $\mathscr{E}E(T) = \bar{\mu}$; equality holds if and only if $T = T_{GDo}$ given by (4.4).*

**Proof.** The requirement $\mathscr{E}E(T) = \bar{\mu}$ is equivalent to the two conditions

$$\Sigma_{\mathscr{S}} p(s) \{w_{0s} + \Sigma_s w_{ks} b_k\} = \bar{b} \tag{4.6}$$

$$\Sigma_{\mathscr{S}} p(s)\, \Sigma_s h_{ks} = 1 \tag{4.7}$$

where $h_{ks} = w_{ks} a_k$. Analyzing $\mathscr{E}E(T - \bar{\mu})^2 = E\mathscr{V}(T) + E\{\mathscr{B}(T)\}^2$, note first that $E\mathscr{V}(T) = \sigma^2 R - (1-\rho)\sigma^2 Q$, where

$$\begin{aligned} R &= \Sigma_{\mathscr{S}} p(s) (\Sigma_s h_{ks})^2 \\ Q &= \Sigma_{\mathscr{S}} p(s) \{(\Sigma_s h_{ks})^2 - \Sigma_s h_{ks}^2\} \end{aligned} \tag{4.8}$$

Using the Cauchy–Schwartz inequality and equation (4.7) we have from (4.8)

$$R \geqslant \frac{\{\Sigma_{\mathscr{S}} p(s)\, \Sigma_s h_{ks}\}^2}{\Sigma_{\mathscr{S}} p(s)} = 1$$

with equality holding if and only if $\Sigma_s h_{ks} = c$, where $c$ is independent of $s$. Next, let us maximize $Q$ subject to (4.7). Let $U = \Sigma_{\mathscr{S}} p(s) \Sigma_s h_{ks}$. Equating $\partial\{Q - 2\lambda(U-1)\}/\partial h_{ks}$ to zero gives $nM$ equations, where $M$ is the number of $s$ for which $p(s) > 0$,

$$2p(s)(\Sigma_s h_{ks} - h_{ks} - \lambda) = 0$$

We obtain $\lambda = (n-1)/n$ and $h_{ks} = 1/n$, that is,

$$w_{ks} = \frac{1}{n a_k}, \quad k \in s, \; s \in \mathscr{S} \tag{4.9}$$

Moreover, $Q \leqslant 1 - 1/n$, and therefore $E\mathscr{V}(T) \geqslant \sigma^2\{1 - (1-\rho)(1-1/n)\}$. In both cases, equality is obtained if and only if (4.9) holds.

We note finally that $E\{\mathscr{B}(T)^2\} \geqslant 0$, where equality is attained in particular if $w_{ks}$ is given by (4.9) and $w_{os} = \Sigma_1^N b_k/N - \Sigma_s b_k/na_k$. These choices of $w_{os}$ and $w_{ks}$ also satisfy the constraint (4.6) which completes the proof. □

We can now easily prove the main result of this section:

**Theorem 4.1.** *Under Model $G_T$, letting $A = \Sigma_1^N a_k^2/N$,*

$$\mathscr{E}V(p,T) \geqslant \mathscr{E}V(p_o, T_{GDo}) = \frac{(1-\rho)(1-fA)\sigma^2}{n}$$

*for any strategy $(p,T)$ such that $p$ is an FES(n) design with $\alpha_k > 0$ $(k = 1, \ldots, N)$, and $T \in \mathscr{L}_u$, the class of all linear (homogeneous or nonhomogeneous) p-unbiased predictors of $\bar{Y}$; equality holds if and only if $(p,T) = (p_o, T_{GDo})$, where $p_o$ and $T_{GDo}$ are given by (4.3) and (4.4), respectively.*

**Proof.** Since $(p,T)$ is $p$-unbiased, formula (3.3) of Lemma 4.1 applies, where $\mathscr{V}(\bar{Y}) = \{\rho + (1-\rho)A/N\}\sigma^2$ is a constant.

The requirement of $p$-unbiasedness is equivalent to a set of $N+1$ conditions given by $\Sigma_{\mathscr{S}} p(s) w_{os} = 0$ and, letting $C_k = \{s: k \in s\}$,

$$\Sigma_{C_k} p(s) w_{ks} = \frac{1}{N} \quad (k = 1, \ldots, N)$$

These $N+1$ conditions imply (4.6) and (4.7), that is, any $p$-unbiased predictor is $p\xi$-unbiased for $\bar{\mu}$. Thus, from Lemma 4.2 it follows that, for any FES($n$) design $p$ such that $(p,T)$ is $p$-unbiased,

$$E\mathscr{E}(T-\bar{\mu})^2 \geqslant E\mathscr{E}(T_{GDo} - \bar{\mu})^2$$

But $E\mathscr{E}(T_{GDo} - \bar{\mu})^2 = \{1 + \rho(n-1)\}\sigma^2/n$ for any design $p$, hence, in particular, for $p = p_o$ defined by (4.3). Therefore, $E\mathscr{E}(T-\bar{\mu})^2 \geqslant \{1+\rho(n-1)\}\sigma^2/n$ for any $p$-unbiased $(p,T)$ allowed by the theorem, and equality holds if and only if $(p,T) = (p_o, T_{GDo})$. □

**Remark 1.** Theorem 4.1 says that it is optimal under Model $G_T$ to use a design giving large inclusion probabilities to units considered by the model to be highly variable, that is, to the units considered to have unpredictable $Y$-values. In the example where the units are farms and $Y$ is acres under wheat, the large farms would frequently be assigned large values $a_k$ when the model is constructed.

However, if all $Y_k$ are considered to have identical model variances, that is, $a_k = 1$ $(k = 1, \ldots, N)$, then $(p_o, T_{Do})$ is the best strategy, where $p_o = p_o(s)$ is such that $\alpha_k = f$ $(k = 1, \ldots, N)$, which is satisfied, for example, for the design *srs*, and

$$T_{Do} = \bar{Y}_S + \bar{b} - \bar{b}_S$$

the traditional difference predictor, where $\bar{b}_S = \Sigma_S b_k/n$. The expected variance is

$$\mathscr{E}V(p_o, T_{Do}) = \frac{(1-\rho)(1-f)\sigma^2}{n}$$ ■

**Remark 2.** Theorem 4.1 applies for the regression model, Model G$_R$, with $\beta$ *known*, as it is a special case of Model G$_T$ such that $\mu = \rho = 0$, $a_k^2 \propto v(x_k)$, $b_k = \beta x_k$ $(k = 1, \ldots, N)$, assuming known auxiliary variable values $x_k$. Let $t_k = \sqrt{v(x_k)}$ $(k = 1, \ldots, N)$ and define $m_{t1} = \Sigma_1^N t_k/N$, and

$$R_{Yt} = \frac{\Sigma_S Y_k/t_k}{\nu(S)}$$

$$R_{xt} = \frac{\Sigma_S x_k/t_k}{\nu(S)}$$

Under the conditions of Theorem 4.1, the minimum $\mathscr{E}V$ strategy is then given by $(p_o, T_{GDo})$, where the inclusion probabilities of the FES($n$) design $p_o$ are $\alpha_k = ft_k/m_{t1}$, assuming $nt_k < Nm_{t1}$ for all $k$, and

$$T_{GDo} = m_{t1}R_{Yt} + \beta(\bar{x} - m_{t1}R_{xt}) \tag{4.10}$$

(In the expressions for $R_{Yt}$ and $R_{xt}$, set $\nu(S) = n$ because of the FES($n$) property.) Moreover, letting $A = \Sigma_1^N t_k^2/Nm_{t1}^2$, the expected variance is $\mathscr{E}V(p_o, T_{GDo}) = (1 - fA)\sigma^2/n$.

However, if $\beta$ is unknown and to be replaced in (4.10) by an estimate $\hat{\beta}$ then the resulting predictor, although in general not $p$-unbiased, will have certain attractive properties to be further investigated in Chapter 7. ■

A strategy based on the $p$-unbiased Horvitz–Thompson predictor $T_{HT}$ given by (4.2) can never, under Model G$_T$, be better than the strategy $(p_o, T_{GDo})$. This fact, expressed in the following corollary of Theorem 4.1, is intuitively obvious, because $T_{HT}$ has fewer ' free' constants to be chosen than $T_{GD}$ given by (4.1).

**Corollary 4.1.** *Under Model G$_T$, and for any choice of positive inclusion probabilities $\alpha_k$ $(k = 1, \ldots, N)$ of an FES(n) design p,*

$$\mathscr{E}V(p, T_{HT}) \geq \mathscr{E}V(p_o, T_{GDo}) = \frac{(1-\rho)(1-fA)\sigma^2}{n}$$

*where equality holds if and only if $p = p_o$ and $b_k \propto a_k$ $(k = 1, \ldots, N)$ in which* *case* $T_{GDo} = T_{HTo}$, *say, where*

$$T_{HTo} = \Sigma_S \frac{Y_k}{n a_k} \tag{4.11}$$

*so that the minimum* $\mathscr{E}V$ *strategy becomes* $(p_o, T_{HTo})$, $p_o$ *being given by (4.3).* □

**Remark 3.** A related result was obtained by Hájek (1959). Assuming that $\mathscr{E}(Y_k) = \mu_k$, $\mathscr{E}(Y_k - \mu_k)^2 = \sigma_k^2$, $\mathscr{E}(Y_k - \mu_k)(Y_l - \mu_l) = 0$ $(k \neq l)$, and considering the class of linear homogeneous $p$-unbiased predictors, that is, $T = \Sigma_S w_{kS} Y_k$, he showed that the minimum $\mathscr{E}V$ strategy requires: (i) $\alpha_k \propto \sigma_k$, (ii) $w_{ks} = 1/N\alpha_k$ for all $k \in s$, and (iii) $N \Sigma_s \mu_k w_{ks} = \Sigma_1^N \mu_k$ for all $s$. ■

**Remark 4.** The following growth model indicates a practical situation in which a $p$-unbiased strategy based on the generalized difference predictor $T_{GDo}$ is likely to outperform any $p$-unbiased strategy involving the Horvitz–Thompson predictor.

In constructing Model $G_T$, suppose we choose $b_k$ as the known value for unit $k$ of the variable of interest at the time of the previous survey. From that time until the time of the present survey, unit $k$ has grown by an unknown amount $y_k - b_k$. Assuming we can accurately designate those units for which growth has been particularly strong, we should assign large values $a_k$ to such units. The units with large $a_k$ are not necessarily the ones for which $b_k$ is large. The low growth units should be assigned small values $a_k$. Let $p_o$ be the design for which the inclusion probabilities $\alpha_k$ are proportional to the assessed growth factors $a_k$. Unless the $a_k$ are roughly proportional to the $b_k$, we can then expect the $p$-unbiased strategy $(p_o, T_{GDo})$ to have a considerably smaller $p$-variance than, say, $(p_1, T_{HT})$, where $T_{HT} = \Sigma_S Y_k/N\alpha_k$ and the inclusion probabilities of $p_1$ satisfy $\alpha_k \propto b_k$ for $k = 1, \ldots, N$. ■

**Example 1.** This example relates the Horvitz–Thompson strategy to the stratified random sampling procedure. Let $V_h$ $(h = 1, \ldots, L)$, be pairwise disjoint label sets, $\cup_{h=1}^{L} V_h = \{1, 2, \ldots, N\}$ and let $N_h$ be the number of labels in $V_h$, $\Sigma_1^L N_h = N$. ($\Sigma_1^L$ will denote summation over $h$ from 1 to $L$.) Consider the special case of Model $G_T$ which specifies common model means and variances within strata, that is, $\mu_k = \mu_h$ and $\sigma_k = \sigma_h$ for all $k \in V_h$ $(h = 1, \ldots, L)$; assume also $b_k \propto a_k$ $(k = 1, \ldots, N)$ and $\rho = 0$.

By Corollary 4.1, the minimum $\mathscr{E}V$ strategy is $(p_o, T_{HTo})$, where $p_o$ is any FES($n$) design having the inclusion probabilities $\alpha_k = f a_k = n\sigma_h / \Sigma_1^L N_h \sigma_h$ for all $k \in V_h$, and

$$T_{HTo} = \frac{(\Sigma_1^L N_h \sigma_h)(\Sigma_1^L T_{Yh}/\sigma_h)}{Nn}$$

where $T_{Yh} = \Sigma_{k \in s_h} Y_k$, $s_h \subseteq V_h$ being the set of $n_h$ labels sampled from $V_h$; $\Sigma_1^L n_h = n$.

For comparison, consider the usual stratified sampling procedure, that is, simple random sampling without replacement of $n_h$ units within each stratum. The inclusion probabilities are

$$\alpha_k = \frac{n_h}{N_h} \qquad (k \in V_h,\ h = 1, \ldots, L)$$

The predictor form of the usual $p$-unbiased estimator is

$$T_{ST} = \Sigma_1^L \frac{N_h T_{Yh}}{N n_h}$$

The expected variance under the assumed model is

$$\mathcal{E} E(T_{ST} - \bar{Y})^2 = \frac{\Sigma_1^L N_h^2 (1/n_h - 1/N_h) \sigma_h^2}{N^2}$$

This is minimized by selecting $n_h$ according to the well known optimum allocation rule,

$$n_h = \frac{n N_h \sigma_h}{\Sigma_1^L N_h \sigma_h} \qquad (h = 1, \ldots, L)$$

If the numbers calculated by this rule happen to be integers then $T_{ST}$ is seen to be identical to $T_{HTo}$. In general, however, the numbers $nN_h\sigma_h/\Sigma_1^L N_h\sigma_h$ will turn out to be non-integer, and rounding off to the nearest integer will give a stratified sampling procedure somewhat less efficient than $(p_o, T_{HTo})$; compare Hájek (1959, p. 410). Thus the optimum allocation stratified sampling procedure is at best as efficient (in the $\mathcal{E}V$ sense) as the Horvitz–Thompson strategy $(p_o, T_{HTo})$.

The stratified sampling procedure capitalizes on the idea that the precision of the estimate is increased by selecting relatively many units from a stratum for which the variance $\sigma_h^2$ is large.

In the Horvitz–Thompson strategy, the same goal is achieved by giving relatively large selection probabilities to the units for which the variance $\sigma_h^2$ is large.

The correspondence just discussed between the Horvitz–Thompson strategy and the stratified sampling strategy is not surprising, since stratification is nothing but a practically expedient way to achieve unequal inclusion probabilities for units not in the same stratum. In this form, classical survey samplers were using unequal probability sampling as a means to reduce variance long before the Horvitz–Thompson (1952) paper gave a different twist to the idea. ■

## 4.5. OPTIMAL DESIGN-UNBIASED STRATEGIES: MODEL $E_T$

In this section we impose the stronger (compared to Model $G_T$) requirement of exchangeability. It is shown that exchangeability ensures the optimality of $(p_o, T_{GDo})$ in the wider class of all $p$-unbiased predictors obtained when the linearity restriction of the previous section is removed.

The label part of the data $d = \{(k, y_k): k \in s\}$ is in itself informationless for estimating $\bar{Y}$. However, we cannot without further justification conclude that $k$ can be dropped in $d$, so that only the unlabeled data $\{y_k : k \in s\}$ are retained. Certain models do, however, provide the formal justification for dropping the label part of $d$, with no loss of information, in an $\mathscr{E}V$ sense. The following Lemma 4.5 deals with this situation. First we need to recall some results in general statistical theory.

Assume that $(X_1, \ldots, X_n)$ has an $n$-dimensional exchangeable distribution $G(x_1 \ldots, x_n)$ belonging to the family $\mathscr{G}$ of $n$-dimensional absolutely continuous exchangeable distributions on the Borel sets of $R_n$. Let $E_G(X_i) = \mu$ be the common mean of the $X_i$ $(i = 1, \ldots, n)$. Let $g(x_1, \ldots, x_n)$ be the joint probability density of $X_1, \ldots, X_n$. Let $X_i = x_i$ denote the outcome of the $i$th $X$-variable, where no two $x_i$ are equal. The set of values, $\{x_1, \ldots, x_n\}$, has probability density

$$n!\, g(x_1, \ldots, x_n) \tag{5.1}$$

since the set obtains under any one of the $n!$ permutations $r_1, \ldots, r_n$ of $1, \ldots, n$ such that $X_{r_i} = x_i$ $(i = 1, \ldots, n)$. In particular, the probability density of the order statistic, $x_{(1)} \leqslant x_{(2)} \leqslant \ldots \leqslant x_{(n)}$, where $x_{(i)}$ denotes the $i$th smallest of $x_1, \ldots, x_n$ is also given by (5.1).

In the special case where $X_1, \ldots, X_n$ are independently and identically distributed with unidimensional distribution function $G(x)$, and density function $g(x)$, then (5.1) reduces to

$$n!\, \Pi_1^n\, g(x_i)$$

The following lemma, due to Fraser (1957, p. 30) and presented here without proof, establishes the completeness of the family of distributions of $\{X_1, \ldots, X_n\}$ when the $X_i$ are independently and identically distributed with common distribution $G \in \mathscr{G}_1$, the class of all absolutely continuous distributions over $R_1$. Let $\mathscr{G}_n$ be the class of all distributions over $R_n$ such that each of the independent $X_i$ have the distribution $G \in \mathscr{G}_1$.

**Lemma 4.3.** *Let $X_1, \ldots, X_n$ be independently and identically distributed with common distribution $G \in \mathscr{G}_1$ and let $h = h(x_1, \ldots, x_n)$ be any real symmetric function. Then $E_{G_n}(h) = 0$ for all $G_n \in \mathscr{G}_n$ implies that $h = 0$ almost everywhere $(G_n)$.* □

Clearly, any symmetric function $h(x_1, \ldots, x_n)$ is a function of the set $\{x_1, \ldots, x_n\}$ and vice versa, any function of the set is symmetric.

There is an equivalent of Lemma 4.3, due to Halmos (see Fraser, 1957, p. 31), in the class of all discrete distributions.

Assume now that $X_1, \ldots, X_n$ have an $n$-dimensional exchangeable distribution $G \in \mathcal{G}$, the class of all $n$-dimensional absolutely continuous exchangeable distributions over $R_n$.

The class $\mathcal{G}_n$ discussed in Lemma 4.3 is contained in $\mathcal{G}$, and the class $\mathcal{G}_n$ was already wide enough to ensure that $E_{G_n}(h) = 0$ implies $h = 0$ almost everywhere. As pointed out by Fraser (1957, p. 26), we can then extend Lemma 4.2 from $\mathcal{G}_n$ to the wider class $\mathcal{G}$:

**Lemma 4.4.** *Let $X_1, \ldots, X_n$ be random variables with the n-dimensional exchangeable distribution $G \in \mathcal{G}$, and let $h = h(x_1, \ldots, x_n)$ be any real symmetric function. Then $E_G(h) = 0$ for all $G \in \mathcal{G}$ implies that $h = 0$ almost everywhere $(G)$.* □

Using Lemma 4.4, we show another preliminary result. Lemma 4.5, concerned with the best $p\xi$-unbiased estimator of $\bar{\mu}$, is the general class analog of Lemma 4.2, which was restricted to the linear class.

**Lemma 4.5.** *Let p be any given FES(n) design with $\alpha_k > 0$ $(k = 1, \ldots, N)$. Then, under Model $E_T$,*

$$\mathscr{E}E(T - \bar{\mu})^2 \geqslant \mathscr{E}E(T_{GDo} - \bar{\mu})^2 = \frac{\{1 + \rho(n-1)\}\sigma^2}{n}$$

*for any (linear or nonlinear) $p\xi$-unbiased estimator T of $\bar{\mu}$; equality holds if and only if $T = T_{GDo}$ given by (4.4).*

**Proof.** Assume without loss of generality that the finite population values $z_k = (y_k - b_k)/a_k (k = 1, \ldots, N)$ are distinct. Under the model, $Z_1, \ldots, Z_N$ have an exchangeable absolutely continuous distribution with density, say, $g(z_1, \ldots, z_N)$. It follows that the marginal density of any subset of $n$ $Z$-variables, say, $Z_{k_1}, \ldots, Z_{k_n}$, is the same, $g(z_{k_1}, \ldots, z_{k_n})$, say.

Consider an outcome $\mathscr{D} = d$, that is, $S = s = \{k_1, \ldots, k_n\}$ and $Z_{k_i} = z_{k_i}$ $(i = 1, \ldots, n)$, say. The probability density of the outcome $\mathscr{D} = d$ is

$$p_{\mathscr{D}}(d) = p(s)g(z_{k_1}, \ldots, z_{k_n}) \tag{5.2}$$

which obtains from (3.1), noting that because of the exchangeability there is no need to keep the $s$ indexing $g$.

We must find $T = T(\mathscr{D})$ to minimize the $p\xi$-variance $\mathscr{E}E(T - \bar{\mu})^2$ subject to $\mathscr{E}E(T) = \bar{\mu} = \mu + \bar{b}$.

Let $Z_S = \{Z_k : k \in S\}$ denote the random variable of which a typical outcome is the set of unlabeled $z$-data $z_s = \{z_k : k \in s\}$, under the combined random mechanism where $S = s$ with probability $p(s)$ and $Z_1, \ldots Z_N$ are distributed with exchangeable joint density $g(z_1, \ldots, z_N)$. Note that the set of numbers $z_s$ may be the outcome of the random variables $Z_k$ associated with any set $s' \in \mathcal{S}$. If $p_{Z_S}(\cdot)$ denotes the density function of $Z_S$, we have

$$p_{Z_S}(z_s) = \sum_{s' \in \mathcal{S}} p_{Z_S}(z_s \mid S = s')\, P(S = s')$$

Let $\Sigma_\pi$ denote summation over the $n!$ permutations of the $n$ labels of any fixed set $s'$ and consider the event $Z_{s'} = z_s$. The symmetry of $g$ gives

$$p_{Z_S}(z_s \mid S = s') = \Sigma_\pi\, g(z_{k_1}, \ldots, z_{k_n}) = n!\, g(z_{k_1}, \ldots, z_{k_n})$$

independently of $s'$, and therefore

$$\begin{aligned} p_{Z_S}(z_s) &= \sum_{s' \in \mathcal{S}} n!\, g(z_{k_1}, \ldots, z_{k_n})\, p(s') \\ &= n!\, g(z_{k_1}, \ldots, z_{k_n}) \end{aligned} \tag{5.3}$$

It follows from (5.2) and (5.3) that the conditional density of $\mathcal{D}$ given $Z_S = z_s$ is $p_{\mathcal{D}}(d \mid z_s) = p(s)/n!$, which is independent of $g(\cdot)$. Therefore, the statistic $Z_S$ is sufficient, that is, for any statistic $T = \phi(\mathcal{D})$, we can find one with no larger value of $\mathscr{E}E(T - \bar{\mu})^2$ that is based on $Z_S$ alone, namely, $\mathscr{E}E\{\phi(\mathcal{D}) \mid Z_S\}$, using the usual argument in Rao-Blackwellization. Thus we can restrict the search to statistics $T$ that depend on $\mathcal{D}$ only through $Z_S$.

Now, the distribution (5.3) of $Z_S$ is the same as (5.1), that is, the same as that of the "order statistic" in making an observation on the random vector $(X_1, \ldots, X_n)$ with exchangeable density $g(x_1, \ldots, x_n)$. By Lemma 4.4, the family of distributions for the order statistic is complete.

Hence, the mean $\bar{Z}_S = \Sigma_S Z_k/n$ is the unique predictor which is a function of $Z_S$ and $p\xi$-unbiased for $\mu = \mathscr{E}E(Z_k)$ $(k = 1, \ldots, N)$. Thus $T_{GDo} = T_{GDo}(\mathcal{D}) = \bar{Z}_S + \bar{b}$ is the minimum $p\xi$-variance $p\xi$-unbiased estimator of $\bar{\mu} = \mu + \bar{b}$. □

We can now easily prove the main result of this section.

**Theorem 4.2.** *Under Model $E_T$, letting $A = \Sigma_1^N a_k^2/N$,*

$$\mathscr{E}\,V(p, T) \geqslant \mathscr{E}\,V(p_o, T_{GDo}) = \frac{(1-\rho)(1-fA)\sigma^2}{n}$$

*for any strategy $(p, T)$ such that $p$ is an FES(n) design with $\alpha_k > 0$ $(k = 1, \ldots, N)$, and $T \in \mathcal{A}_u$, the class of all (linear or nonlinear) p-unbiased predictors of $\bar{Y}$; equality holds if and only if $(p, T) = (p_o, T_{GDo})$, where $p_o$ and $T_{GDo}$ are given by (4.3) and (4.4) respectively.*

**Proof.** Similarly as in the proof of Theorem 4.1, note that formula (3.3) of Lemma 4.1 applies, with $\mathscr{V}(\bar{Y}) = \{\rho + (1-\rho)A/N\}\sigma^2$, a constant. Now, $T$ is $p$-unbiased for $\bar{Y}$ and therefore $p\xi$-unbiased for $\mathscr{E}(\bar{Y}) = \bar{\mu} = \mu + \bar{b}$.

By Lemma 4.5, $\mathscr{E}E(T-\bar{\mu})^2$ is minimized, for any given design $p$ covered by the theorem, by $T = T_{GDo}$, and $E\mathscr{E}(T_{GDo} - \bar{u})^2 = \{1 + \rho(n-1)\}\sigma^2/n$. The latter equation holds in particular for $p = p_o$ given by (4.3), and $T_{GDo}$ is $p$-unbiased under $p_o$. Therefore, $E\mathscr{E}(T-\bar{\mu})^2 \geqslant \{1+\rho(n-1)\}\sigma^2/n$ for any $p$-unbiased strategy $(p, T)$, with equality if and only if $(p, T) = (p_o, T_{GDo})$. □

**Remark 1.** Theorem 4.2 can be shown to hold under the class of discrete exchangeable distributions $\xi$ covered by the Random permutation model, Model $\mathrm{E}_{RP}$ (see Table 4.1). That is, $(p_o, \mathrm{T}_{GDo})$ is the best strategy, and

$$\mathscr{E}V(p_o, T_{GDo}) = \frac{\{N/(N-1)\}(1-fA)\sigma_z^2}{n}$$

where $\sigma_z^2$ is given by (2.2). ■

**Remark 2.** Under the special case $\mathrm{E}_{T0}$ of Model $\mathrm{E}_T$ (see Table 4.1), the minimum $\mathscr{E}V$ strategy is, according to Theorem 4.2, $(p_o, \bar{Y}_S)$, where $p_o$ is an FES($n$) design with inclusion probabilities $\alpha_k = f$ $(k = 1, \ldots, N)$, for example, *srs*, and $\bar{Y}_S = (\Sigma_S Y_k)/n$, the sample mean. Under random labeling, Model $\mathrm{E}_{RP0}$, this was shown by Thompson (1971) in generalization of a result by C. R. Rao (1971) restricted to linearity. In other words, when the labels are uninformative with respect to the $y_k$ (in the sense implied by Model $\mathrm{E}_{T0}$, that is, the $Y_k$ are exchangeable), then a $p$-unbiased procedure based on the sample mean is optimal, and hence better in this situation than all other $p$-unbiased procedures, including those of Horvitz–Thompson, Murthy, and Lahiri–Midzuno–Sen. Needless to say, these latter procedures are more efficient than $(p_o, \bar{Y}_s)$ under models making use of auxiliary information, see Chapter 7. Due to the basic importance of the sample mean, we return to a discussion of this predictor in Section 5.7. ■

We proceed to a discussion of the Horvitz–Thompson predictor under Model $\mathrm{E}_T$. Just as under Model $\mathrm{G}_T$ (see Corollary 4.1) the strategy $(p, T_{HT})$ can never be more efficient in an $\mathscr{E}V$ sense than $(p_o, T_{GDo})$. However, if $b_k \propto a_k$ $(k = 1, \ldots, N)$, then $T_{GDo}$ reduces to $T_{HTo}$ given by (4.11). Thus, we have the following corollary of Theorem 4.2.

**Corollary 4.2.** *Under the special case of Model* $\mathrm{E}_T$ *where* $b_k \propto a_k$ $(k = 1, \ldots, N)$,

$$\mathscr{E}V(p, T) \geqslant \mathscr{E}V(p_o, T_{HTo})$$

*for any strategy* $(p, T)$ *such that* $p$ *is an FES($n$) design with* $\alpha_k > 0$

*$(k = 1, \ldots, N)$ and $T \in \mathcal{A}_u$; equality holds if and only if $(p, T) = (p_o, T_{HTo})$, where $p_o$ and $T_{HTo}$ are given by (4.3) and (4.11), respectively.* □

**Remark 3.** The result of Corollary 4.2 was obtained by Godambe and Thompson (1973) by first showing the statement to hold under Model $\mathrm{E}_{RP}$ with $b_k = 0$ $(k = 1, \ldots, N)$, and generalizing from there. They also provided an example to show that the statement is violated if $p$ is a non-FES design. ■

We present a simple numerical example to further illustrate the Horvitz–Thompson strategy under random labeling.

**Example 1.** Let $N = 3$ and $n = 2$. We formalize our prior knowledge about the unknown $\mathbf{y}$ by assuming Model $\mathrm{E}_{RP}$ to hold with the $z$-values 2.4, 4, 8 and with the labeled numbers $a_1 = 3/4, a_2 = 1, a_3 = 5/4$, and $b_1 = b_2 = b_3 = 0$.

The design giving probabilities $p_o(s_1) = 1/6, p_o(s_2) = 1/3, p_o(s_3) = 1/2$ to the samples $s_1 = \{1, 2\}$, $s_2 = \{1, 3\}$, $s_3 = \{2, 3\}$ has the inclusion probabilities $\alpha_k = fa_k$ $(k = 1, 2, 3)$, where $a_k$ are as specified above and $f = 2/3$. This design combined with the estimator $T_{HTo}$ given by (4.11) constitutes a minimum $\mathscr{E}V$ strategy under the model.

As a competing $p$-unbiased strategy, consider the design *srs*, that is, $p(s_i) = 1/3$ $(i = 1, 2, 3)$, combined with $\bar{Y}_S = \Sigma_S Y_k/2$. In an $\mathscr{E}V$ sense, this strategy is less efficient, as confirmed by calculation of the $\mathscr{E}V$'s: $\mathscr{E}V(p_o, T_{HTo}) = 1.27$; $\mathscr{E}V(srs, \bar{Y}_S) = 1.66$.

Under the model, each of six population vectors $\mathbf{y}$ is assigned the probability 1/6, namely, each vector $\mathbf{y} = (y_1, y_2, y_3)$ such that $(4y_1/3, y_2, 4y_3/5)$ equals any one of the six vectors formed by arranging the $z$-values 2.4, 4, 8 in a particular order.

One of these populations is $\mathbf{y} = (6, 4, 3)$. If for this fixed population we consider $p$-variance (not $\xi$-expected $p$-variance) as the criterion, then the strategy $(srs, \bar{Y}_S)$ is strongly preferable to $(p_o, T_{HTo})$, the $p$-variances being 0.39 and 1.36, respectively. Hence the order of preference between the two strategies is reversed, compared with the expected variance criterion computed above.

The reason is that $(p_o, T_{HTo})$ is a very poor strategy, from a $p$-variance point of view, for the population $\mathbf{y} = (6, 4, 3)$, because the inclusion probabilities of $p_o$, that is, $\alpha_1 = 3/6$, $\alpha_2 = 4/6$, $\alpha_3 = 5/6$, are inversely related to the $y$-values $y_1 = 6$, $y_2 = 4$, $y_3 = 3$. (A more dramatic illustration of this is Basu's (1971) amusing "Circus Example".)

On the other hand, another of the six populations is $\mathbf{y} = (1.8, 4, 10)$. Here the $p$-variance of $(p_o, T_{HTo})$ would be considerably less than that of $(srs, \bar{Y})$ because of the positive correlation between $y_k$ and $\alpha_k$.

Moreover, we know that the *average* $p$-variance over all six populations is the smallest, among all $p$-unbiased strategies, for the choice $(p_o, T_{HTo})$. ■

**Remark 4.** In spite of their optimality properties, the strategies $(p_o, T_{GDo})$ and $(p_o, T_{HTo})$ are not free of weaknesses.

**Table 4.2. Some Optimal $p$-Unbiased Strategies for FES($n$) Designs under the $\mathscr{E}\mathscr{V}$ Criterion.**

| Model | Optimal Strategy: Design | Optimal Strategy: Predictor | Expected Variance |
|---|---|---|---|
| $G_T$, $E_T$, or $E_{RP}$ | $\alpha_k = fa_k$ | $T_{GDo} = \Sigma_S(Y_k - b_k)/na_k + \bar{b}$ (generalized difference predictor) | $(1-\rho)(1-fA)\sigma^2/n$ |
| $G_R$ with $\beta$ known | $\alpha_k = ft_k/m_{t1}$ | $T_{GDo} = m_{t1}R_{Yt} + \beta(\bar{x} - m_{t1}R_{xt})$ (generalized difference predictor) | $(1-fA)\sigma^2/n$ |
| $G_T$, $E_T$, or $E_{RP}$ with all $a_k = 1$ | $\alpha_k = f$ | $T_{Do} = \bar{Y}_S + \bar{b} - \bar{b}_S$ (difference predictor) | $(1-\rho)(1-f)\sigma^2/n$ |
| $G_T$, $E_T$, or $E_{RP}$ with $b_k \propto a_k$ | $\alpha_k = fa_k$ | $T_{HTo} = \Sigma_S Y_k/na_k$ (Horvitz–Thompson predictor) | $(1-\rho)(1-fA)\sigma^2/n$ |
| $G_{T0}$, $E_{T0}$, or $E_{RP0}$ | $\alpha_k = f$ | $\bar{Y}_S = \Sigma_S Y_k/n$ (sample mean) | $(1-\rho)(1-f)\sigma^2/n$ |

*Note.* Under Model $E_{RP}$, $\rho = -1/(N-1)$ and $\sigma^2 = \sigma_z^2$. For detailed description of each model, see Table 4.1. Under G-models, optimality applies within classes of linear predictors; under E-models, there is no restriction to linearity.

Consider for example the situation where the means of several $y$-variables have to be estimated, using PPS sampling with inclusion probabilities proportional to known positive values $x_1, \ldots, x_N$ of one single auxiliary variable.

If $T_{HT}$ is used for estimating all of the $y$-means, the variance will be large for those $y$-variables having a weak or negative correlation with $x$.

J. N. K. Rao (1966a) has suggested that $p$-biased alternatives should be entertained to cope with the multicharacter problem in PPS sampling. He suggested alternative predictors with MSE smaller than the variance of the usual $p$-unbiased predictors, and with a bias that is small relative to the standard error, under conditions where $x$ and $y$ are essentially unrelated.

He showed, for example, that the $p$-biased alternative predictor $\bar{Y}_S = \Sigma_S y_k/n$ has smaller $\mathscr{E}$MSE than the $p$-unbiased predictor $T_{HT} = \Sigma_S y_k/N\alpha_k$, when the design is such that $\alpha_k \propto x_k$ for $k = 1, \ldots, N$, and for a model that expresses unrelatedness of $y$ and $x$, namely, $\mathscr{E}(Y_k) = \mu$, $\mathscr{V}(Y_k) = \sigma^2$ for $k = 1, \ldots, N$ and $\mathscr{C}(Y_k, Y_l) = 0$ for all $k \neq l$. ∎

We conclude by giving in tabular form a summary of results following from Sections 4.4–4.5. Table 4.2 deals with strategies involving well-known and widely used predictors, the generalized difference estimator being perhaps an exception. It is of interest to note that for each of these predictors, $T_{GDo}$, $T_{HTo}$, $\bar{Y}_S$, and $T_{Do}$, there is a superpopulation model which grants the status of minimum $\mathscr{E}V$ $p$-unbiased strategy. The first line of Table 4.2 summarizes the statements of Theorems 4.1 and 4.2. The subsequent lines are corollaries of those two theorems.

CHAPTER 5

# Inference under Superpopulation Models: Prediction Approach using Tools of Classical Inference

In Chapters 2–4, statistical inference about $\bar{y} = \Sigma_1^N y_k/N$ was guided by the requirement of $p$-unbiasedness. The man-made randomization imposed by the statistician through his choice of sampling design was essential in determining the optimal strategy.

By contrast, the approaches considered in Chapters 5 and 6 are justified solely by an assumed model. Design-related considerations are pushed into the background; $p$-unbiasedness is no longer considered necessary. Such a point of departure is contrary to much of classical thought in survey sampling. However, reliance on a realistic and valid superpopulation model can give powerful inferences. Traditionalists in survey sampling feel hesitant about the approach, one reason being that they are unwilling to give up the idea that randomization through sampling design is essential. In addition, they would argue that grossly invalid inferences would result if the assumed model were unrealistic. Nevertheless, the model approach appears to be of great potential value for survey sampling practice.

## 5.1. PREDICTING THE POPULATION MEAN

The data $d = \{(k, y_k) : k \in s\}$ form the starting point for the inference, as in previous chapters, but the outlook is different: The sample is given, and what remains for the statistician to do is to predict the values of $y_k$ for the $N - \nu(s)$ unobserved coordinates of $\mathbf{y} = (y_1, \ldots, y_N)$. This will lead to a prediction of $\bar{y} = \Sigma_1^N y_k/N$. The superpopulation may be indexed by an unknown parameter $\boldsymbol{\theta}$, so that the prediction of $y_k$, for $k \notin s$, may involve estimation, in the classical or Bayesian sense, of the unknown $\boldsymbol{\theta}$, through the observed $y_k$. Consequently, as pointed out by Kalbfleisch and Sprott (1969), the approaches of this chapter and the next bring survey sampling theory into close resemblance with ordinary statistical theory, whether it be "classical," as in the present chapter, or "Bayesian" or "fiducial" as in Chapter 6. The tools of point estimation, including the use of likelihood, will play the same role as in traditional statistical theory, in that they serve to estimate the parameters of the hypothesized superpopulation, the ultimate goal being to predict the finite population mean.

Consider an arbitrary set $s \in \mathscr{S}$, containing $\nu(s)$ labels. Let $\tilde{s}$ denote the set of labels not in $s$, that is, $\tilde{s} = \{1, \ldots, N\} - s$.

Consider a class of superpopulation distributions (that is, a class of probability measures on $R_N$), $\xi = \xi_\theta$, where $\xi$ is the distribution of $Y_1, \ldots, Y_N$ and where the indexing parameter $\boldsymbol{\theta}$ is usually unknown, $\boldsymbol{\theta} \in \Theta$.

Denote by $\xi_s = \xi_{s;\theta}$ the marginal distribution of $Y_{k_1}, \ldots, Y_{k_{\nu(s)}}$, where $k_1 < \cdots < k_{\nu(s)}$ is an enumeration in increasing order of the labels $k \in s$. Denote by $\xi_{\tilde{s}|s} = \xi_{\tilde{s}|s;\theta}$ the conditional distribution of $Y_k$ for $k \in \tilde{s}$, taken in the order of increasing $k$, given $Y_{k_1}, \ldots, Y_{k_{\nu(s)}}$.

In cases where density functions are involved, let $g(\mathbf{y} \mid \boldsymbol{\theta})$ be the density function of $Y_1, \ldots, Y_N$, where $\mathbf{y} = (y_1, \ldots, y_N)$.

Let $g_s(\mathbf{y}_s \mid \boldsymbol{\theta})$ be the marginal density function of $Y_{k_1}, \ldots, Y_{k_{\nu(s)}}$, where $\mathbf{y}_s = (y_{k_1}, \ldots, y_{k_{\nu(s)}})$.

Let $g_{\tilde{s}|s}(\mathbf{y}_{\tilde{s}} \mid \boldsymbol{\theta}) = g(\mathbf{y} \mid \boldsymbol{\theta})/g_s(\mathbf{y}_s \mid \boldsymbol{\theta})$ be the conditional density function of $Y_k$, $k \in \tilde{s}$, in order of increasing $k$, given $Y_{k_1}, \ldots, Y_{k_{\nu(s)}}$.

Note that if $\xi$ is an exchangeable distribution, then $\xi_s$ and $\xi_{\tilde{s}|s}$ are exchangeable. Under exchangeability, the functions $g_s(\cdot \mid \boldsymbol{\theta})$ are identical for all $s$ containing $\nu(s)$ labels, and so are the functions $g_{\tilde{s}|s}(\cdot \mid \boldsymbol{\theta})$, so that the model expresses the same prior opinion on the $y$-values associated with any one of the sets $s$, regardless of which $\nu(s)$ labels make up the set; in this sense, labels are uninformative.

Denote by $\mathscr{E}$, $\mathscr{E}_s$, and $\mathscr{E}_{\tilde{s}|s}$, respectively, the expectation operators associated with $\xi$, $\xi_s$, and $\xi_{\tilde{s}|s}$, that is, with $g(\cdot \mid \boldsymbol{\theta})$, $g_s(\cdot \mid \boldsymbol{\theta})$, and $g_{\tilde{s}|s}(\cdot \mid \boldsymbol{\theta})$.

As in Chapter 4, the minimum $\mathscr{E}$MSE criterion will be considered. Unless otherwise stated, all designs $p$ to be considered are assumed to be noninformative as defined in Section 1.4.

The following preliminary analysis of the criterion will benefit both the classical approach of this chapter and the Bayesian approach of Chapter 6:

The $\mathscr{E}$MSE of an arbitrary strategy $(p, T)$ is

$$\mathscr{E}\text{MSE}(p, T) = \mathscr{E}E(T - \bar{Y})^2$$

If $p$ is noninformative, the operators $\mathscr{E}$ and $E$ may be interchanged, that is,

$$\mathscr{E}\text{MSE}(p, T) = E\mathscr{E}(T - \bar{Y})^2 = E\mathscr{E}_s\mathscr{E}_{\tilde{s}|s}(T - \bar{Y})^2 \tag{1.1}$$

In the model-based approaches of Chapters 5–6, a natural objective is to choose

$T$ to minimize $\mathscr{E}(T - \bar{Y})^2$ for any sample $s$ that might have resulted. Hence, if we can find $T^*$ to minimize $\mathscr{E}(T - \bar{Y})^2$ for any $s \in \mathscr{S}$, and $p$ is noninformative, then $T^*$ also has the property of minimizing $\mathscr{E}\text{MSE}(p, T)$ for any given design $p$.

Thus $T^*$, the minimum $\mathscr{E}$MSE predictor, will be independent of a noninformative design.

However, if we wish to proceed one step further and consider the presampling question of finding the best design to use with the predictor $T^*$, then $\mathscr{E}\text{MSE}(p, T^*)$ should be minimized for variations in $p$.

For any given $s \in \mathscr{S}$, we can write

$$\bar{Y} = f_s \bar{Y}_s + (1 - f_s)\bar{Y}_{\bar{s}} \tag{1.2}$$

where $\bar{Y}_s = (\Sigma_s Y_k)/\nu(s)$; $\bar{Y}_{\bar{s}} = (\Sigma_{\bar{s}} Y_k)/\{N - \nu(s)\}$, and $f_s = \nu(s)/N$. In terms of the actual finite population vector $(y_1, \ldots, y_N)$, an outcome of $(Y_1, \ldots, Y_N)$, (1.2) can be written as

$$\bar{y} = f_s \bar{y}_s + (1 - f_s)\bar{y}_{\bar{s}}$$

Basu (1971) argued that in this representation of $\bar{y}$, the sample mean $\bar{y}_s$ is known, and that the statistician should, therefore, attempt a post-survey prediction of the mean $\bar{y}_{\bar{s}}$ of the nonsurveyed units.

While admitting that a decision-theorist might object to making the choice of estimator after looking at the data, Basu (1971) nevertheless considered such an approach to represent the "heart of the matter" in estimating the finite population mean.

In Chapter 4, the superpopulation was used for studying $p$-unbiased strategies. By contrast, in Chapters 5–6 we go to a full-fledged "model-based approach," or "prediction approach," as it is also called.

Brewer (1963b) obtained some early results in this approach, which was then developed in further detail by Royall (1970a, 1971a), Royall and Herson (1973a,b).

In the prediction approach, the superpopulation distribution $\xi$ plays an essential role. In the classical inference version considered in this chapter, the observed $y_k$ are first used to make inference about a possibly unknown parameter vector $\boldsymbol{\theta}$ indexing $\xi$. Finally, $\xi$ is used to predict the mean $\bar{Y}_{\bar{s}}$ of the unobserved units.

The predictive argument, in somewhat different forms, is also found in the fiducial approach of Kalbfleisch and Sprott (1969) and in the Bayesian approach of Ericson (1969a); see Chapter 6.

These approaches attach little or no importance to the way in which the sample was selected. A consequence would be, for example, that we need not worry about using sophisticated PPS sampling plans; indeed, Basu (1971) maintains that "it is not easy to understand how surveyors got messed up with the idea of unequal probability sampling." Royall (1970a) merely asserts that "the sample $s$ should be one which permits a good predictor (of $Y_{\bar{s}}$) to be constructed." Kalbfleisch and Sprott (1969) feel, however, that inference based on a superpopulation model "does not contradict the intuitive concept of selecting a sample at random" and that, in fact, simple random sampling (possibly within strata) should be used in selecting $s$. Royall and Herson's (1973a,b) balanced sampling idea (see Section 7.7) is another indication that proponents of the model-based approach really favor "representative" (rather than "extreme") samples as an insurance against errors in the model. Even though an inference can be made for both kinds of sample, one would have more faith in the one based on the representative sample.

In this chapter we shall apply the classical inference variety of the prediction approach to a number of the models reviewed in Table 4.1. Among interesting applications not specifically dealt with here, we mention the use of time series analysis for estimation in repeated surveys; see Blight and Scott (1973), Scott and Smith (1974).

Consider any predictor $T$ of $\bar{Y}$; it can be represented, for any given sample $s$, as

$$T = f_s \bar{Y}_s + (1 - f_s)U \tag{1.3}$$

where $U$ is considered a predictor of $\bar{Y}_{\bar{s}}$. If $f_s = 1$ for some $s \in \mathscr{S}$, define $(1 - f_s)U$ as 0. Note that $T$ becomes $\bar{Y}$, the target of our prediction, if $\nu(S) = N$ with probability one, that is, if the whole population is observed. Since $s$ is given, the statistics $U = U(d)$ and $T(d) = f_s \bar{Y}_s + (1 - f_s)U(d)$ derive their stochastic structure entirely from $\xi$. (Recall from Section 4.3 that $d = \{(k, Y_k) : k \in s\}$.)

We can now rewrite (1.1) in terms of $U$, thereby obtaining

$$\mathscr{E}E(T - \bar{Y})^2 = E\{(1 - f_S)^2 \mathscr{E}_S \mathscr{E}_{\bar{S}|S}(U - \bar{Y}_{\bar{S}})^2\} \tag{1.4}$$

If $\xi$ is completely known, it is clear that minimum $\mathscr{E}$MSE obtains if, for every given $s$, we choose

$$U = \mathscr{E}_{\bar{s}|s}(\bar{Y}_{\bar{s}}) \tag{1.5}$$

This conclusion will be of importance in Chapter 6, where tools of Bayesian estimation theory will be used.

If, however, $\xi$ depends on an unknown parameter vector $\boldsymbol{\theta}$, this $\boldsymbol{\theta}$ must first be estimated. The following representation of the $\mathscr{E}$MSE will then be useful, assuming that $U$ is $\xi$-unbiased for $\bar{Y}_{\bar{s}}$:

$$\mathscr{E}E(T - \bar{Y})^2 = E[(1 - f_S)^2\{\mathscr{V}(U) + \mathscr{V}(\bar{Y}_{\bar{S}}) - 2\mathscr{C}(U, \bar{Y}_{\bar{S}})\}] \tag{1.6}$$

Tools of classical statistical estimation theory will be useful in finding a suitable predictor $U$ of $\bar{Y}_{\bar{s}}$, whereupon $T$ will be determined by (1.3).

Note that if the predictors $T$ and $U$ are related through (1.3) then the following holds:

$T$ is $\xi$-unbiased as a predictor of $\bar{Y}$ (or as an estimator of $\bar{\mu}$) if and only if, for every $s \in \mathscr{S}$, $U$ is $\xi$-unbiased as a predictor of $\bar{Y}_{\bar{s}}$ (or as an estimator of $\bar{\mu}_{\bar{s}}$).

## 5.2. SOME RESULTS ON OPTIMAL $\xi$-UNBIASED PREDICTION

In Theorems 4.1 and 4.2 we showed that the design $p_o$, combined with the generalized difference predictor $T_{GDo}$, given respectively by formulas (4.3) and (4.4) of Section 4.4, gives an optimal $p$-unbiased strategy. We show below that the predictor $T_{GDo}$ can be improved upon, for any *fixed* design, if $\xi$-unbiasedness, but not $p$-unbiasedness, is required.

Theorem 5.1 derives, for any given design, the minimum $\mathscr{E}$MSE linear $\xi$-unbiased predictor under Model $G_T$. Theorem 5.3 is a more specialized result requiring the superpopulation distribution to be such that $Y_1, \ldots, Y_N$ are independent, as in several of the models introduced in Section 4.2.

**Theorem 5.1.** *Let p be any given design. Then, under Model* $G_T$,

$$\mathscr{E}E(T - \bar{Y})^2 \geqslant \mathscr{E}E(T^* - \bar{Y})^2$$

*where T is any linear $\xi$-unbiased predictor of $\bar{Y}$, and, for any $s \in \mathscr{S}$,*

$$T^* = f_s\bar{Y}_s + (1 - f_s)(\bar{Z}_s\bar{a}_{\bar{s}} + \bar{b}_{\bar{s}}) \tag{2.1}$$

*where*

$$\bar{Z}_s = \frac{\Sigma_s Z_k}{\nu(s)} \tag{2.2}$$

*and $Z_k = (Y_k - b_k)/a_k$ $(k = 1, \ldots, N)$; equality holds if and only if $T = T^*$.*

**Proof.** If a linear and $\xi$-unbiased predictor $T$ is of the form (1.3), then $U$ must have the form $U = w_{0s} + \Sigma_s w_{ks} Y_k$. Using that $\mathscr{E}(Y_k) = \mu_k = \mu a_k + b_k$, we obtain

$$\mathscr{E}(U) = \bar{\mu}_{\bar{s}} = \mu\bar{a}_{\bar{s}} + \bar{b}_{\bar{s}} \tag{2.3}$$

where $\bar{\mu}_{\bar{s}} = \Sigma_{\bar{s}}\mu_k/\{N - \nu(s)\}$ and $\bar{a}_{\bar{s}}$, $\bar{b}_{\bar{s}}$ are analogously defined. From (2.3) it follows that

$$\Sigma_s w_{ks} a_k = \bar{a}_{\bar{s}} = \Sigma_{\bar{s}}\, a_k/\{N - \nu(s)\}$$

which in turn implies $\mathscr{C}(U, \bar{Y}_{\bar{s}}) = \rho\sigma^2\bar{a}_{\bar{s}}$, independently of the choice of the $w_{ks}$. Therefore, the $\mathscr{E}$MSE given by (1.6) is minimized if, for every fixed $s \in \mathscr{S}$, we choose $U$ to minimize $\mathscr{V}(U)$, subject to (2.3), where $\mu$ is the only unknown quantity.

But by generalized least squares theory, the uniformly minimum $\xi$-variance linear ξ-unbiased estimator of $\mu$ is $\bar{Z}_s$, because under Model $G_T$ the random variables $Z_1, \ldots, Z_N$ have identical first- and second-order moments. Hence in order to minimize $\mathscr{V}(U)$ we should choose $U = U^* = \bar{Z}_{\bar{s}}\bar{a}_{\bar{s}} + \bar{b}_{\bar{s}}$. This means that $T = T^*$ will minimize the criterion $\mathscr{E}\,E(T - Y)^2$. □

In Theorems 4.1 and 4.2 we were concerned with the predesign stage choice of the entire strategy, design plus predictor. By contrast, in Theorem 5.1, a fixed design is assumed, and ξ-unbiased predictors are compared.

Defining $T^\circ = \bar{Z}_s + \bar{b}$, we can express $T^*$ in terms of $T^\circ$ plus a remainder term,

$$T^* = T^\circ + \frac{\Sigma_s(a_k - \bar{a}_s)Z_k}{N}$$

Note that $T^\circ = T_{GDo}$ given by (4.4) of Section 4.4 in the special case where $p$ is an FES($n$) design. Especially if the sampling fraction is small, the value of the remainder term is likely to be small, which leaves little difference between $T^*$ and $T^\circ$

**Remark 1.** In various interesting special cases of Model $G_T$, the predictor $T^*$ given by (2.1) can be expressed as follows:

(i) Under Model $G_T$ with all $b_k = 0$:

$$T^* = \bar{Z}_s + \frac{\Sigma_s(a_k - \bar{a}_s)Z_k}{N}$$

where $\bar{Z}_s = \Sigma_s Z_k / \nu(s)$ and $Z_k = Y_k / a_k$. If $p$ is an FES($n$) design, then

$$T^* = T_{HTo} + \frac{\Sigma_s(a_k - \bar{a}_s)Z_k}{N} \tag{2.4}$$

where $T_{HTo}$ is given by (4.11) of Section 4.4.

(ii) Under Model $G_T$ with all $a_k = 1$:

$$T^* = T_{Do} = \bar{Y}_s + \bar{b} - \bar{b}_s$$

the difference predictor.

(iii) Under Model $G_{To}$:

$$T^* = \bar{Y}_s$$

the sample mean. ■

For the sake of comparison let us further investigate the intuitively appealing predictor of $\bar{Y}$ given by

$$T^\circ = \bar{Z}_s + \bar{b} \tag{2.5}$$

where $\bar{Z}_s$ is given by (2.2). It has the following properties:

(i) $T^\circ$ is a $\xi$-unbiased predictor of $\bar{Y}$ under Model $G_T$;

(ii) $T^\circ$ minimizes, under Model $G_T$, and for any fixed $s$, the criterion $\mathscr{E}(T - \bar{\mu})^2$, among linear $\xi$-unbiased estimators of the superpopulation parameter $\bar{\mu} = \mu + \bar{b}$; hence $T^\circ$ would be the preferred predictor if inference were directly to the superpopulation and not to the realization $y_1, \ldots, y_N$;

(iii) if $p$ is an FES($n$) design, $T^\circ = T_{GDo}$ given by (4.4) of Section 4.4;

(iv) if all $b_k = 0$ and $p$ is an FES($n$) design, $T^\circ = T_{HTo}$ given by (4.11) of Section 4.4.

Thus, $T^*$ given by (2.1) as well as $T^\circ$ given by (2.5) are optimal, but by different criteria. In terms of our principal criterion, minimum $\mathscr{E}$ MSE, $T^*$ is optimal and better than $T^\circ$ to an extent shown in Theorem 5.2, of which the straightforward algebra proof is omitted.

**Theorem 5.2.** *Under Model $G_T$, for any design $p$,*

$$\mathscr{E}E(T^\circ - \bar{Y})^2 - \mathscr{E}E(T^* - \bar{Y})^2 = \frac{E\{\Sigma_S(a_k - \bar{a}_S)^2\}(1-\rho)\sigma^2}{N^2} \geqslant 0 \tag{2.6}$$

*where $T^*$ and $T^\circ$ are given by (2.1) and (2.5) respectively. Moreover,*

$$\mathscr{E}E(T^\circ - \bar{Y})^2 = \left[ E\left\{\frac{1}{\nu(S)} - \frac{A}{N}\right\} + \frac{2E(A - \bar{a}_S)}{N}\right](1-\rho)\sigma^2 \tag{2.7}$$

*and $\bar{a}_s = \Sigma_s a_k/\nu(s)$, and $A = \Sigma_1^N a_k^2/N$.*

*Strict inequality holds in (2.6) if $p(s) > 0$ for some $s$ such that not all $a_k$ for $k \in s$ are equal.* □

**Remark 2.** The comparison of Theorem 5.2 holds for any $p$. Thus neither $T^*$ nor $T^\circ$ are necessarily $p$-unbiased; both are, however, $\xi$-unbiased under the model.

If $p$ is any FES($n$) design, then Theorem 5.2 implies that $(p, T^*)$ is at least as good a strategy as $(p, T_{GDo})$. This holds in particular for $(p_o, T^*)$ when compared to the $p$-unbiased strategy $(p_o, T_{GDo})$, where $p_o$ is given by (4.3) of Section 4.4. Under the design $p_o$, $E(A - \bar{a}_S)$ in (2.7) vanishes, leaving $\mathscr{E}E(T_{GDo} - \bar{Y})^2 = (1-\rho)(1 - fA)\sigma^2/n$.

Also, under Model $G_T$ with all $b_k = 0$, $(p, T^*)$ is at least as good a strategy as $(p, T_{HTo})$, for any FES($n$) design $p$; in this case $T^*$ is given by (2.4).

The expression (2.6) in the case $p = p_o$ is interesting in that it represents the "price" we have to pay, in terms of increased $\mathscr{E}$MSE, if we wish to keep the property of $p$-unbiasedness. This magnitude of (2.6) tends to be small when $N$ is large and/or $f = n/N$ is small. ■

We examine next the problem of optimal $\xi$-unbiased prediction in the simpler situation where $Y_1, \ldots, Y_N$ are *independently* distributed, as in Models $G_{PI}$, $E_{PI}$, $G_R$, and $G_{MR}$; see Table 4.1. The following theorem will have wide applicability in the following sections. Different versions of the theorem appear in Fuller (1970), Royall (1970a).

**Theorem 5.3.** *Let $p$ be any given design, and let $T = f_s\bar{Y}_s + (1 - f_s)U$ and $T' = f_s\bar{Y}_s + (1 - f_s)U'$ be $\xi$-unbiased predictors of $\bar{Y}$.*

*Then, if $\xi$ is a product measure (as under Models $G_{PI}$, $E_{PI}$, $G_R$, and $G_{MR}$), the inequality*

$$\mathscr{E}E(T - \bar{Y})^2 \leqslant \mathscr{E}E(T' - \bar{Y})^2$$

*holds if and only if, for every $s \in \mathscr{S}$ such that $p(s) > 0$,*

$$\mathscr{V}(U) = \mathscr{E}(U - \bar{\mu}_{\bar{s}})^2 \leqslant \mathscr{V}(U') = \mathscr{E}(U' - \bar{\mu}_{\bar{s}})^2$$

*If for some $s$ with $p(s) > 0$ the latter inequality is strict, then the former inequality is also strict.*

**Proof.** The $\mathscr{E}$MSE of any predictor $T$ expressed by (1.3) is given by (1.4). Since $\xi$ is a product measure, the term $\mathscr{C}(U, \bar{Y}_{\bar{s}})$ vanishes, leaving

$$\mathscr{E}E(T - \bar{Y})^2 = E[(1 - f_S)^2\{\mathscr{V}(U) + \mathscr{V}(\bar{Y}_{\bar{S}})\}] \tag{2.8}$$

which proves the theorem. □

**Remark 3.** An obvious implication of Theorem 5.3 is the following: Let $\omega$ denote a specified class of $\xi$-unbiased estimators of $\bar{\mu}_{\bar{s}}$ in the sense of classical statistical inference theory. To $\omega$ corresponds an implied class $\tau$ of predictors $T = f_s\bar{Y}_s + (1 - f_s)U$ such that $T \in \tau$ if and only if $U \in \omega$. Clearly, $\tau$ consists of $\xi$-unbiased predictors of $\bar{Y}$. For example, if $\omega$ is a linear class, $\tau$ will also be composed of linear predictors $T$.

By Theorem 5.3, the minimum $\mathscr{E}$MSE predictor in $\tau$, for any fixed $p$, is given by $T_o = f_s\bar{Y}_s + (1 - f_s)U_o$, where $U_o$ has minimum $\xi$-variance among $U \in \omega$. To find $U_o$ is strictly a problem of classical statistical inference theory. $T_o$ also has the property of minimizing $\mathscr{E}(T - \bar{Y})^2$ for any given $s$, and for $T \in \tau$. ■

**Remark 4.** In line with the comment made in Section 5.1, the optimal predictors in Theorems 5.1 and 5.3 are independent of the design $p$. If we wish to seek the best design for the optimal predictors, then the $\mathscr{E}$MSE will have to be minimized for variations in $p$. An illustration of this is given in Section 5.4. ■

## 5.3. PREDICTION WITHOUT AUXILIARY VARIABLE INFORMATION

Consider Model $E_{PI}$, that is, $\xi$ belongs to a class of product measures, the $Y_k$ having a common distribution, and hence common means $\mu$ and variances $\sigma^2$.

Let $G(y \mid \boldsymbol{\theta})$, or simply $G(y)$, denote the distribution function of each $Y_k$; let $g(y \mid \boldsymbol{\theta})$, or simply $g(y)$, be the density function. Depending on what is known about $g(\cdot \mid \boldsymbol{\theta})$, the class of distributions $\xi$ may be further specified, as in the following three cases considered by Wretman (1970):

*Case A:* $G(y)$ belongs to the class of absolutely continuous distributions, with unknown mean $\mu = \int_{-\infty}^{\infty} y \, dG(y)$ and unknown variance $\sigma^2 = \int_{-\infty}^{\infty} (y-\mu)^2 dG(y)$. The shape $G(\cdot)$ is also unknown.

*Case B:* $G(y) = G(y \mid \boldsymbol{\theta})$ is the class of all absolutely continuous distributions of the form $G(y \mid \boldsymbol{\theta}) = G_0\{(y - \theta_1)/\theta_2\}$, where $G_0(\cdot)$ is a known function ("known shape"), but where the location parameter $\theta_1$ and the scale parameter $\theta_2$ are unknown. The model implies that the $Y_k$ have the common unknown mean, $\mu = \theta_1 + c_1\theta_2$, and the common unknown variance, $\sigma^2 = (c_2 - c_1^2)\theta_2^2$, where $c_r = \int_{-\infty}^{\infty} z^r dG_0(z)$ $(r = 1, 2)$ are known constants.

*Case C:* $G(y)$ is completely known; in particular the mean $\mu$ and the variance $\sigma^2$ are known.

The three cases represent increasing levels of knowledge about the superpopulation. Cases A and B correspond to models which may often be realistically assumed, Case C is included mostly for the sake of comparison since the complete knowledge it assumes is practically never available. Starting with C, the simplest case, we make the following inferences about $\bar{Y}$:

*CASE C.* The criterion (1.4) is minimized if, for every $s \in \mathscr{S}$, $U = \mathscr{E}_{\bar{s}|s}(\bar{Y}_{\bar{s}}) = \mu$, which is known. The minimum $\mathscr{E}$MSE predictor is, from (1.3),

$$T = f_s\bar{Y}_s + (1 - f_s)\mu$$

with $\mathscr{E}$MSE given by

$$\mathscr{E}\,\text{MSE}(p, T) = \frac{E(1 - f_S)\sigma^2}{N}$$

*CASE A.* The inference in this situation is summarized by the following theorem:

**Theorem 5.4.** *Let p be any design. Then, under the conditions of Model $E_{PI}$ and Case A,*

$$\mathscr{E}E(T - \bar{Y})^2 \geqslant \mathscr{E}E(\bar{Y}_S - \bar{Y})^2 = E\left\{\frac{1}{\nu(S)} - \frac{1}{N}\right\}\sigma^2$$

*where T is any $\xi$-unbiased predictor of $\bar{Y}$; equality holds if and only if $T = \bar{Y}_s$.*

**Proof.** Applying Theorem 5.3, we must find, for any fixed $s$, the minimum $\xi$-variance predictor $U = U(d)$ of $\bar{Y}_{\bar{s}}$ such that $\mathscr{E}(U) = \mathscr{E}(Y_k) = \mu$, where $d = \{(k, Y_k): k \in s\}$.

The search for such a predictor $U(\mathscr{d})$ can be restricted to those that are symmetric in $Y_k$ for $k \in s$, that is, to functions $U = U(Y_s)$, where $Y_s = \{Y_k: k \in s\}$ and the $\nu(s)$ variables $Y_k$ are independently and identically distributed with common mean $\mu$. (This can be shown in a manner similar to the proof of Lemma 3.6: For any nonsymmetric $\xi$-unbiased $U = U(\mathscr{d})$, we can construct one with smaller $\xi$-variance that depends on $\mathscr{d}$ through $Y_s$ only.)

Now, $U = Y_s$ is $\xi$-unbiased for $\mu$ and is a symmetric function of the $Y_k$ for the given sample $s$. By Lemma 4.3, $\bar{Y}_s$ is unique as a symmetric $\xi$-unbiased estimator of $\mu$. Hence $U = \bar{Y}_s$ has minimum $\xi$-variance among predictors $U = U(\mathscr{d})$ such that $\mathscr{E}(U) = \mu$. By Theorem 5.3, $T = T(\mathscr{d}) = f_s\bar{Y}_s + (1 - f_s)\bar{Y}_s = \bar{Y}_s$ is then the desired optimal predictor. □

**Remark 1.** By Remark 1 of Section 5.2, $\bar{Y}_s$ is, for an arbitrary FES($n$) design, the minimum $\mathscr{E}$MSE *linear* $\xi$-unbiased predictor of $\bar{Y}$ under Model $G_{TO}$ (or Model $G_{PI}$). Thus, Theorem 5.4 removes the linearity restriction, at the price of requiring exchangeability; compare Table 5.1 below. ■

*CASE B.* When the distribution $G(y) = G_0\{(y - \theta_1)/\theta_2\}$ is known except for $\theta_1$ and/or $\theta_2$, classical parametric estimation theory may be used in predicting $\bar{Y}$.

The added information (*vis-à-vis* Case A) of known superpopulation shape will often make it possible to improve, in the sense of reducing the $\mathscr{E}$MSE, on the predictor $T = \bar{Y}_s$ found to be best in Case A.

From (2.8), provided $\mathscr{E}(U) = \mu$,

$$\mathscr{E}E(T - \bar{Y})^2 = E\{(1 - f_S)^2\, \mathscr{V}(U) + (1 - f_S)\sigma^2/N\}$$

If the conditions of the Cramér–Rao inequality are satisfied with respect to the family of distributions at hand, we know a lower bound on $\mathscr{V}(U)$:

Let $g(y \mid \theta)$, where $\theta \in \Theta$, define a one-parameter family $\mathscr{G}_\theta$ of distributions. Let

$$\mu = \mu(\theta) = \int_{-\infty}^{\infty} yg(y \mid \theta)\, dy$$

and

$$C_\theta = \frac{\{\mu'(\theta)\}^2}{\mathscr{E}\left\{\dfrac{\partial \ln g_s(\mathbf{y}_s \mid \theta)}{\partial\theta}\right\}^2}$$

where $g_s(\mathbf{y}_s \mid \theta) = \Pi_s g(y_k \mid \theta)$ and $\mu'(\theta) = d\mu(\theta)/d\theta$. The Cramér–Rao inequality, provided it is applicable, states that

$$\mathscr{V}(U) \geqslant C_\theta \tag{3.1}$$

It is known (see Wijsman, 1973) that the lower bound $C_\theta$ can be attained if and only if $\mathscr{G}_\theta$ is a one-parameter exponential family of distributions.

**Example 1.** Consider the family $\mathscr{G}_\theta$ of gamma distributions such that $g(y \mid \theta) = \{1/\theta^r \Gamma(r)\} y^{r-1} \exp(-y/\theta)$ for $y > 0$, where $\theta > 0$ is unknown and $r$ is known. We have $\mu(\theta) = r\theta$, $\sigma^2(\theta) = \int_0^\infty (y - r\theta)^2 g(y \mid \theta)\, dy = r\theta^2$, and $C_\theta = \sigma^2(\theta)/\nu(s)$. The inequality (3.1) reduces in this example to an equality for the choice $U = \bar{Y}_s$. Hence, by Theorem 4.1, $T = f_s \bar{Y}_s + (1 - f_s)\bar{Y}_s = \bar{Y}_s$ is the minimum $\mathscr{E}$MSE predictor of $\bar{Y}$, uniformly in $\mathscr{G}_\theta$, and regardless of the design $p$. ■

**Example 2.** Consider the one-parameter family $\mathscr{G}_\theta$ of normal distributions such that

$$g(y \mid \theta) = (1/\sqrt{2\pi}\,\sigma)\exp\{-(x - \theta)^2/2\sigma^2\}$$

$-\infty < \theta < \infty$, $\theta$ unknown and $\sigma^2$ known. The minimum $\mathscr{E}$MSE predictor of $\bar{Y}$ is again $T = \bar{Y}_s$, uniformly in $\mathscr{G}_\theta$, and regardless of $p$. ■

However, for most other shapes, it is possible to derive $\xi$-unbiased predictors $T$ that perform better in an $\mathscr{E}$MSE sense than $\bar{Y}_s$. Here classical estimation theory based on order statistics is helpful.

Let, for a given $s$ of size $\nu(s)$, $U$ belong to the class of $\xi$-unbiased estimators of $\mu$ that are linear homogeneous in the order statistics $Y_{(1)} \leqslant \cdots \leqslant Y_{(\nu(s))}$, that is,

$$U = \sum_{i=1}^{\nu(s)} g_i Y_{(i)} \tag{3.2}$$

Lloyd (1952), using the Gauss–Markov theorem, presented the theoretical background for computing the optimal coefficients $g_i = g_{oi}$, say. "Optimal" here means that the $g_{oi}$ give minimum $\xi$-variance among estimators $U$ that are $\xi$-unbiased for $\mu$ and linear in the order statistics $Y_{(i)}$, given the shape and the sample size $\nu(s)$. Tables of the optimal coefficients $g_{oi}$ have been computed for a wide range of distribution shapes, see Sarhan and Greenberg (1962), and the references in David (1970).

If $U$ is of the form (3.2), the implied predictor of $\bar{Y}$, that is, $T = f_s \bar{Y}_s + (1 - f_s)U$, is also linear in the order statistics. The minimum $\mathscr{E}$MSE predictor $T$ in this implied class is, for any $p$, given by

$$T_o = \sum_{i=1}^{\nu(s)} g_i' Y_{(i)}$$

where $g_i' = 1/N + (1 - f_s) g_{oi}$ $(i = 1, \ldots, \nu(s))$. Thus computation of the weights $g_i'$ is easy in the cases where the $g_{oi}$ can be obtained from existing tables, for example, the tables referred to above.

The gain in efficiency realized by $(p, T_o)$ over $(p, \bar{Y}_s)$ may be expressed as

$$G = \frac{\mathscr{E}\,\mathrm{MSE}(p, \bar{Y}_S)}{\mathscr{E}\,\mathrm{MSE}(p, T_o)} - 1$$

which depends on $N, n, p$ and the shape. In the case of an arbitrary FES($n$) design, computations by Wretman (1970) showed that, for $N = \infty$, $n = 5$, the value of $G$ is 39% for the rectangular shape, 27% for the Laplace shape. Gains in efficiency can thus be considerable, if the additional knowledge of the superpopulation assumed by Case B is at hand.

## 5.4. PREDICTION USING AUXILIARY INFORMATION: MODEL $G_R$

We now consider the frequently used regression model, Model $G_R$, assuming that $x_k > 0$ ($k = 1, \ldots, N$) are known auxiliary variable values.

The main result of the section, Theorem 5.5, is due to Brewer (1963b) and Royall (1970a). The theorem derives, for an arbitrary design, the minimum $\mathscr{E}$MSE linear homogeneous $\xi$-unbiased (to be abbreviated $\xi$-BLU) predictor of $\bar{Y}$, to be denoted $T_{BR}$. While the $\xi$-BLU predictor does not depend explicitly on the design, its $\mathscr{E}$MSE does. The question of choice of design is discussed. It turns out that, under a broad range of conditions, it is optimal to select purposively the $n$ units corresponding to the $n$ largest $x$-values.

**Theorem 5.5.** *Under Model $G_R$, and for known auxiliary variable measurements $x_k > 0$ ($k = 1, \ldots, N$), the $\xi$-BLU predictor of $\bar{Y}$ is, for any design $p$, given by*

$$T_{BR} = f_s \bar{Y}_s + (1 - f_s)\hat{\beta}\bar{x}_{\bar{s}} \tag{4.1}$$

*where* $\bar{x}_{\bar{s}} = \Sigma_{\bar{s}}\, x_k / \{N - \nu(s)\}$ *and*

$$\hat{\beta} = \Sigma_s \frac{x_k Y_k}{v(x_k)} \Big/ \Sigma_s \frac{x_k^2}{v(x_k)} \tag{4.2}$$

*Furthermore,*

$$\mathscr{E}\text{MSE}(p, T_{BR}) = \frac{E\{(\Sigma_{\bar{s}} x_k)^2 \mathscr{V}(\hat{\beta}) + \sigma^2 \Sigma_{\bar{s}} v(x_k)\}}{N^2} \tag{4.3}$$

*where*

$$\mathscr{V}(\hat{\beta}) = \frac{\sigma^2}{\Sigma_s x_k^2 / v(x_k)}$$

**Proof.** Appealing to Theorem 5.3, we need to find $U$ such that $\mathscr{E}(U) = \bar{\mu}_{\bar{s}} = \beta\bar{x}_{\bar{s}}$, where $\beta$ is the only unknown quantity, and such that $\mathscr{V}(U)$ is minimized among linear estimators $U$.

By generalized least squares theory, the minimum $\xi$-variance linear $\xi$-unbiased estimator of $\beta$ is, for a given $s$, given by $\hat{\beta}$, formula (4.2). Hence, $T_{BR}$ given by formula (4.1) obtains from Theorem 5.3. Finally, (4.3) follows from (2.8). □

Let us denote by $T_{BRg}$ the predictor $T_{BR}$ obtained if $v(x) = x^g$.

**Remark 1.** If $v(x) = x$, (4.1) reduces to the classical ratio predictor (cf. Basu, 1971, p. 215); that is, $T_{BR1} = T_R$, where

$$T_R = \frac{\bar{x}\bar{Y}_s}{\bar{x}_s} \tag{4.4}$$

with

$$\mathscr{E}\,\mathrm{MSE}(p, T_R) = \frac{\bar{x}E\{N\bar{x}/\Sigma_S x_k - 1\}\sigma^2}{N} \tag{4.5}$$

The predictor $T_R$ is $\xi$-unbiased under Model $G_R$ but can also be $p$-unbiased, namely, for the Lahiri–Midzuno–Sen design, see Section 7.3. ■

**Remark 2.** If $v(x) = x^2$, (4.1) can be rewritten as

$$T_{BR2} = \bar{x}R_{Yx} + f_s(\bar{Y}_s - \bar{x}_s R_{Yx}) \tag{4.6}$$

where

$$R_{Yx} = \frac{\Sigma_s Y_k/x_k}{\nu(s)} \tag{4.7}$$

If $p$ is an FES($n$) design, so that $\nu(s) = n$ in (4.7), then the first term of (4.6) becomes identical to the Horvitz–Thompson predictor corresponding to the design *ppsx* defined in Section 1.4: Setting $\alpha_k = nx_k/N\bar{x}$ in (4.2) of Section 4.4 we get $T_{HT} = \bar{x}R_{Yx}$. Thus $T_{BR2}$ differs by little from the Horvitz–Thompson predictor when the sampling fraction is small. Remark 5 below shows a design that makes $T_{BR2}$ $p$-unbiased. Note also that the predictor $T^*$ given by (2.1), which is the minimum $\mathscr{E}$MSE linear $\xi$-unbiased predictor for a given design $p$ under Model $G_T$, is identical to $T_{BR2}$ if we let $a_k = x_k/\bar{x}$ and $b_k = 0$ ($k = 1, \ldots, N$) in Model $G_T$. ■

**Remark 3.** Royall (1970a) showed that if (i) $p$ is any FES($n$) design, (ii) $nx_k/N\bar{x} < 1$ for $k = 1, \ldots, N$, (iii) $v(x)/x^2$ is a nonincreasing function, then, under Model $G_R$,

$$\mathscr{E}\mathrm{MSE}(p, \bar{x}R_{Yx}) \geqslant \mathscr{E}\mathrm{MSE}(p, T_{BR2})$$

where $T_{BR2}$ is given by (4.6). In particular, this holds for the design *ppsx*, in which case the Horvitz–Thompson strategy (*ppsx*, $\bar{x}R_{Yx}$) is $p$-unbiased whereas (*ppsx*, $T_{BR2}$) is not. (Both $\bar{x}R_{Yx}$ and $T_{BR2}$ are, of course, $\xi$-unbiased under the model). ■

The $\mathscr{E}$ MSE of the *strategy* $(p, T_{BR})$ depends on $p$ through $E(\cdot)$ in (4.3). A before-sampling judgment may sometimes be required as to how $p$ should be chosen in order to minimize $\mathscr{E}\text{MSE}(p, T_{BR})$. A conservative (but not optimal) approach would be to select $s$ with simple random sampling. However, much more efficient ways of selecting $s$ exist, if reliance on the model permits almost complete disregard of the way in which $s$ was obtained. Many sampling practioners would argue that estimates based on units with large $x$-values tend to be "closer" to the population mean than those based on units with small $x$-values, as long as there is a fair amount of positive correlation between $x$ and $y$; compare the discussion in Brewer (1963b) and Royall (1970a). Such a claim receives theoretical support by the following theorem.

Under the fairly broad conditions of Theorem 5.6 below, it turns out that the optimal strategy is to combine $T_{BR}$ with the purposive design that selects with probability one the sample whose $n$ units have the largest $x$-values.

Let $\mathscr{S}_n = \{s : \nu(s) = n\}$ and denote by $s^*$ the set of labels such that

$$\max_{s \in \mathscr{S}_n} \Sigma_s x_k = \Sigma_{s^*} x_k$$

Introducing also $p^* = p^*(s)$ such that

$$p^*(s) = \begin{cases} 1 & \text{if } s = s^* \\ 0 & \text{if } s \neq s^* \end{cases} \tag{4.8}$$

the theorem, due to Royall (1970a), may be formulated as follows.

**Theorem 5.6.** *Let $p$ be any FES($n$) design, and let $p^*$ be defined by (4.8). If $v(x)$ is nondecreasing and $v(x)/x^2$ is nonincreasing, then, under Model $G_R$,*

$$\mathscr{E}\text{MSE}(p, T) \geqslant \mathscr{E}\text{MSE}(p^*, T_{BR})$$

*where $T$ is any linear $\xi$-unbiased predictor of $\overline{Y}$, and $T_{BR}$ is given by (4.1).*

**Proof.** The expression within curly brackets in (4.3) is, under the stated conditions on $v(x)$, minimized if $s = s^*$. Hence, $\mathscr{E}\text{MSE}(p^*, T_{BR}) \leqslant \mathscr{E}\text{MSE}(p, T_{BR})$. The theorem follows by noting also that $\mathscr{E}\text{MSE}(p, T_{BR}) \leqslant \mathscr{E}\text{MSE}(p, T)$, because $T_{BR}$ has minimum $\mathscr{E}$MSE among linear $\xi$-unbiased $T$, regardless of $p$. □

**Remark 4.** The result of Theorem 5.6 sounds extreme and is in conflict with the principles of any statistician convinced of the importance of man-made randomization in form of a design $p$. No $p$-variance can be calculated for the strategy $(p^*, T_{BR})$.

J. N. K. Rao (1975) points out that there are no doubt situations in which the extreme design $p^*$ can be highly efficient for the estimation of *one* $y$-mean, but it is not likely to work well if several $y$-means have to be estimated in the same survey. (The same criticism applies to the common practice of stratification with near 100% selection in the top stratum.) Thus a preferred course of action in the multipurpose study might be simple random sampling and use of auxiliary information *only* at the estimation stage, for example, in the form of a ratio estimator.

Common practice with respect to the ratio predictor $T_R$ is to use it with simple random sampling or with stratified random sampling. The argument leading to the purposive selection of the $n$ units with the largest $x$-values relies heavily on the assumed model holding true. (Compare the discussion in Brewer, 1963b.)

A similar word of caution is expressed by Madow and Hansen (1975). They argue that purposive selection could put the sampler in a difficult spot if it is decided after the survey that some other design should have been used. By contrast, if sufficient randomization has taken place, there is more freedom to do a post-survey re-evaluation of the results, for example, by post-stratification.

If Model $G_R$ were *false*, for example, if $\mathscr{E}(Y_k) = \beta x_k^m$, where $m \neq 1$, then the $\xi$-bias of $T_R$ is

$$\mathscr{E}(T_R - \bar{Y}) = \bar{x}\beta\left(\frac{\Sigma_s x_k^m}{\Sigma_s x_k} - \frac{\Sigma_1^N x_k^m}{\Sigma_1^N x_k}\right) \tag{4.9}$$

While (4.9) is likely to be small under *srs*, it is also clear that an extreme design, such as $p^*$, is likely to produce a considerable $\xi$-bias. Further discussion of $\xi$-bias will be given in Section 7.4. ■

In the philosophy of the model-based approach, $p$-unbiasedness is a property without much interest; $\xi$-unbiasedness is more essential. The following two remarks compare $p$-unbiased reasoning to model-based reasoning.

**Remark 5.** Suppose that the problem is to estimate several $y$-means in the same survey, and that for each $y$-variable we have access to known auxiliary variable values $x_1, \ldots, x_N$. Moreover let $a_1, \ldots, a_N$ be an arbitrary set of known and fixed numbers satisfying $\Sigma_1^N a_k = N$.

Hájek (1959) suggests two predictors, $T_1$ and $T_2$ below, for use in the situation where the $y_k$ and the $x_k$ have a high positive correlation, but where the $y_k$ and the $a_k$ are not necessarily highly correlated. Let $R_{Yx}$ be given by (4.7) and set

$$R_{Ya} = \frac{\Sigma_s Y_k/a_k}{\nu(s)}$$

$$R_{xa} = \frac{\Sigma_s x_k/a_k}{\nu(s)}$$

Assume that the sample size is fixed at $n$ and define

$$T_1 = \bar{x}\{f\bar{Y}_s + (1 - \pi)R_{Ya}\}/\{f\bar{x}_s + (1 - \pi)R_{xa}\} \tag{4.10}$$

and

$$T_2 = \bar{x}R_{Yx} + f(\bar{Y}_s - \bar{x}_s R_{Yx}) + \left\{\frac{(1-\pi)n}{(n-1)}\right\}(R_{Ya} - R_{Yx}R_{xa}) \quad (4.11)$$

where $f = n/N$, $\pi = \Sigma_s a_k/N$, and $\nu(s) = n$ in $R_{Yx}$, $R_{Ya}$, and $R_{xa}$.

Under Model $G_R$, both $T_1$ and $T_2$ are $\xi$-unbiased predictors for $\bar{Y}$, whatever be the numbers $a_1, \ldots, a_N$. By Theorem 5.5 both predictors have greater $\mathscr{E}$MSE than $T_{BR}$, under Model $G_R$ and for any FES($n$) design.

Moreover, both $T_1$ and $T_2$ are related to $T_{BR2}$: If we choose $a_k = x_k/\bar{x}$ for $k = 1, \ldots, N$, then $T_1 = T_2 = T_{BR2}$, provided $f_s = f = n/N$ in (4.6).

As a special case of $T_2$ we obtain the so-called Hartley–Ross estimator (see Section 7.3), namely, if we set $a_k = 1$ for $k = 1, \ldots, N$.

However, in deriving these predictors, Hájek (1959) was motivated by design oriented aspects and $p$-unbiasedness:

Consider an FES($n$) design based on rejective sampling, in which $n$ draws are made one by one with replacement and using probability $p_k = a_k/N$ of drawing label $k$ in each draw. If an already drawn label is repeated, all the selected labels are rejected. The procedure starts all over again and continues until a set $s$ of $n$ distinct labels have been obtained. An advantage is that the same set of $a_k$-values can serve efficiently for the estimation of several $y$-variable means, even though the correlation between the $y_k$ and the $a_k$ may be weak.

Hájek (1959) showed that $T_2$ is $p$-unbiased for $\bar{Y}$ under this rejective sampling design, whereas $T_1$ has a slight $p$-bias for the same design. Since $T_{BR2}$ given by (4.6) is identical to $T_2$ for the choice $a_k = x_k/\bar{x}$, we conclude that $T_{BR2}$ is also $p$-unbiased for this particular case of the rejective design. In general, little is known about the detailed behavior of $T_1$ and $T_2$. ■

**Remark 6.** Finally let us compare $T_{BR}$ with the generalized difference predictor $T_{GDo}$ given by (4.10) of Section 4.4. As long as $\beta$ is known, $(p_o, T_{GDo})$ is the optimal $p$-unbiased strategy. If $\beta$ is unknown and replaced by the estimator $\hat{\beta} = \{\Sigma_s x_k Y_k/t_k^2\}/\{\Sigma_s x_k^2/t_k^2\}$, then we get the predictor

$$T_{GREo} = m_{t1}R_{Yt} + \hat{\beta}(\bar{x} - m_{t1}R_{xt})$$

suggested by Cassel, Särndal, and Wretman (1976). It is a special case of a *generalized regression predictor* $T_{GREG}$ to be considered in Chapter 7. As is easily seen, $T_{GREo}$ is $\xi$-unbiased under Model $G_R$. However, the argument underlying $T_{GREo}$ is design oriented. In contrast to $T_{GDo}$, the predictor $T_{GREo}$ is $p$-biased under the design $p_o$ such that $\alpha_k \propto t_k = \sqrt{v(x_k)}$ for $k = 1, \ldots, N$. The $p$-bias of $(p_o, T_{GREo})$ is small if the regression is linear through the origin.

It follows from Theorem 5.5 that $\mathscr{E}$MSE $(p, T_{BR}) \leqslant \mathscr{E}$MSE $(p, T_{GREo})$ for any fixed design $p$ common to the two strategies. However, a strategy based on

$T_{BR}$ is not necessarily optimal if the idea of an extreme, purposive design, such as $p^*$ given by (4.8), is rejected. The information inherent in Model $G_R$ is put to efficient use already at the design stage if we choose the design $p_o$.

In fact, it will be shown in Remark 4 of Section 7.6 that the design oriented strategy $(p_o, T_{GREo})$ has, under certain conditions, considerably smaller $\mathscr{E}$MSE than the model-based strategy $(p, T_{BR})$ when $p$ is restricted to non-extreme designs such as simple random sampling or balanced sampling, that is, designs prone to producing "representative" samples. (Balanced sampling will be further discussed in Section 7.7 as a device to protect a model-based inference against certain errors in the assumed model.)

This brings into focus some basic questions in the comparison between the design oriented approach and the model-based approach: How much faith can be placed in the model-based inference if the model happens to be invalid? Is it wise to ignore sampling design completely? Should the design not be taken into account at least to the extent where we would have some assurance of drawing "representative" samples, and perhaps even obtaining an approximatively $p$-unbiased estimator, thereby protecting ourselves against the possibility of an invalid model?

No conclusive answers seem available, but we can point to strengths and weaknesses in either argument. Additional discussion is found in this chapter and in Chapter 7, in particular, Section 7.7. ■

## 5.5. JUDGING THE UNCERTAINTY OF THE ESTIMATES

An important methodological question concerns the most appropriate way to measure the uncertainty of an estimate obtained through the Superpopulation approach. Clearly more than one possibility exists. Following Royall (1971a) we discuss the question in terms of an example: Consider the ratio predictor $T_R$ given by (4.4), assume the design to be FES($n$), and assume that the bivariate scatter of $(x_k, y_k)$ $(k = 1, \ldots, N)$ conforms approximately to Model $G_R$ with $v(x) = x$.

In the Fixed population approach, $y_1, \ldots, y_N$ are considered fixed numbers. The conventional measure of uncertainty of the approximately $p$-unbiased estimator $t_R = \bar{x}\bar{y}_s/\bar{x}_s$ is in terms of its $p$-MSE,

$$E(t_R - \bar{y})^2 = \Sigma_{\mathscr{S}}\, p(s)(t_R - \bar{y})^2$$

It is well known from, for example, Cochran (1963, p. 158), that, under the design *srs*, and if $n$ is large,

$$E(t_R - \bar{y})^2 \simeq \frac{(1-f)V}{n} = h_0^2$$

say, where

$$V = \frac{\Sigma_1^N (y_k - \bar{y} x_k / \bar{x})^2}{N - 1}$$

The customary estimate of $V$ is, letting $\hat{\beta} = \Sigma_s y_k / \Sigma_s x_k$,

$$\hat{V} = \frac{\Sigma_s (y_k - \hat{\beta} x_k)^2}{n - 1}$$

so that the estimated $p$-MSE of $t_R$ becomes

$$\frac{(1 - f)\hat{V}}{n} = h_1^2$$

say. Under the assumption $\mathscr{E}(y_k \mid x_k) \propto \mathscr{V}(y_k \mid x_k) \propto x_k$, a simple analysis shows that $h_1^2$ will tend to be large, thus indicating greater uncertainty of the estimate, when a sample containing relatively many units with large $x_k$-values happens to have been selected. Conversely, $h_1^2$ will tend to be small, indicating a highly accurate estimate, if the selected sample happens to be one with predominantly small $x_k$-values. Yet many a sampling practitioner feels that this estimate is "closer" to the population mean in the former case than in the latter.

A third uncertainty measure is the $\xi$-expected $p$-MSE. From (4.5), assuming a FES($n$) design, this quantity is

$$\mathscr{E}E(T_R - \bar{Y})^2 = \frac{\bar{x} E(\bar{x}_{\bar{s}} / \bar{x}_s)(1 - f)\sigma^2}{n}$$

We remarked earlier that in the model-based approach, this quantity is of interest as an uncertainty indicator mainly before a sample has been drawn. *After* the sample has been observed, one may be more inclined to look at the $\xi$-expectation of the squared deviation, $(T - \bar{Y})^2$ , that is, the average value of $(T_R - \bar{Y})^2$ across the different populations that the superpopulation process may yield. Given the sample $s$, a fourth possible uncertainty measure would thus be

$$\mathscr{E}(T_R - \bar{Y})^2 = \frac{(\bar{x} \bar{x}_{\bar{s}} / \bar{x}_s)(1 - f)\sigma^2}{n} \qquad (5.1)$$

In order to be practically useful, the unknown $\sigma$ in (5.1) must first be estimated from the sample. A suggested $\xi$-unbiased estimator of $\sigma^2$ based on weighted squared differences (in view of the heteroscedasticity pattern $v(x) = x$) would be

$$\hat{\sigma}^2 = \frac{\Sigma_s (y_k - \hat{\beta} x_k)^2 / x_k}{n - 1}$$

Hence from (5.1) we have the following uncertainty measure for $t_R$,

$$\frac{(\bar{x} \bar{x}_{\bar{s}} / \bar{x}_s)(1 - f)\hat{\sigma}^2}{n} = h_2^2$$

say. Note that $h_2^2$ tends to be small if the sample $s$ happens to include relatively many units with large $x$-values, and, vice versa, $h_2^2$ is large when many small $x$-value units are included. Thus the tendency of $h_2^2$ is opposite to that of $h_1^2$.

Using $h_1^2$ and $h_2^2$, two possible $100(1-\alpha)\%$ confidence intervals for $\bar{y} = \Sigma_1^N y_k/N$ are

$$t_R \pm z_{\alpha/2} h_1 \tag{5.2}$$

and

$$t_R \pm z_{\alpha/2} h_2 \tag{5.3}$$

where $z_{\alpha/2}$ is standard normal score above which $\alpha/2$ of the probability mass is contained.

As pointed out by Royall (1971a), we may, by appealing to the central limit theorem, claim both to be approximately valid $100(1-\alpha)\%$ confidence intervals for $\bar{y}$, but for completely different reasons.

In the case of the interval (5.2), the same, fixed population is assumed to be sampled repeatedly. Then, under some conditions on $N$, $n$, and the values $(y_1, x_1), \ldots, (y_N, x_N)$, samples $s$ such that the interval (5.2) contains $\bar{y} = \Sigma_1^N y_k/N$ would be selected with the relative frequency of (approximately) $1-\alpha$, assuming the repeated selection of samples $s$ is by the design *srs*.

In interpreting the interval (5.3), the essential ingredient is the probability density ascribed by the superpopulation measure $\xi$ to the various actual populations $y_1, \ldots, y_N$. For a given $s$, and under some conditions on $N$, $n$, and $\xi$, the accumulated $\xi$-probability over all populations such that (5.3) contains the population mean, is approximately $1-\alpha$.

Royall (1971a) argues in line with the model-based approach that (5.3) is the preferable interval. He performed an empirical study involving a number of actual finite populations. For each population he considered the extreme samples consisting of (a) the $n$ units corresponding to the $n$ largest $x$-values; (b) the $n$ units corresponding to the $n$ smallest $x$-values. For the "high" samples, $h_1 > h_0 > h_2$ tended to hold, and for the "low" samples the trend was reversed, confirming the theoretical argument above.

## 5.6. PREDICTION USING AUXILIARY INFORMATION: MODEL $G_{MR}$

The $\xi$-BLU prediction approach can be extended to the multiple regression model, Model $G_{MR}$. Results in this direction have been presented by Royall (1975) and Thomsen (1974). Several examples are given below and the question of maximum likelihood estimation is also considered.

In conformity with the definition of Model $G_{MR}$ given in Section 4.2, we introduce the following notation:

$$\begin{aligned} &\mathscr{E}(\mathbf{Y}_s) = \mathbf{X}_s \boldsymbol{\beta}, \qquad \mathscr{E}(\mathbf{Y}_{\bar{s}}) = \mathbf{X}_{\bar{s}} \boldsymbol{\beta} \\ &\mathscr{V}(\mathbf{Y}_s) = \sigma^2 \boldsymbol{\Sigma}_s, \qquad \mathscr{V}(\mathbf{Y}_{\bar{s}}) = \sigma^2 \boldsymbol{\Sigma}_{\bar{s}}, \qquad \mathscr{C}(\mathbf{Y}_s, \mathbf{Y}_{\bar{s}}) = 0 \end{aligned} \tag{6.1}$$

where $\mathbf{Y}_s$ and $\mathbf{Y}_{\bar{s}}$ are column vectors. The $\nu(s)$ components of $\mathbf{Y}_s$ are $Y_k, k \in s$, and the $N - \nu(s)$ components of $\mathbf{Y}_{\bar{s}}$ are $Y_k$, $k \in \bar{s}$; in both cases the $Y_k$ are enumerated in order of increasing $k$.

Moreover, $\boldsymbol{\beta}' = (\beta_1, \ldots, \beta_q)$ is a vector of unknown parameters. The known matrices $\mathbf{X}_s$ and $\mathbf{X}_{\bar{s}}$ are of order $\nu(s) \times q$ and $\{N - \nu(s)\} \times q$, respectively. The row vector of $\mathbf{X}_s$ or of $\mathbf{X}_{\bar{s}}$ corresponding to unit $k$, $\mathbf{x}_k' = (x_{k1}, x_{k2}, \ldots, x_{kq})$, is such that $x_{k1} = 1$ whatever be the value of $k$ and $x_{k2}, \ldots, x_{kq}$ are the measurements for unit $k$ on $q - 1$ auxiliary variables $x_2, \ldots, x_q$. The diagonal matrices $\boldsymbol{\Sigma}_s$ and $\boldsymbol{\Sigma}_{\bar{s}}$ are $\nu(s) \times \nu(s)$ and $\{N - \nu(s)\} \times \{N - \nu(s)\}$, respectively. The diagonal element for unit $k$ is $v_k$, a known quantity. Hence $\mathscr{V}(Y_k) = \sigma^2 v_k$, where $\sigma^2$ is unknown.

The following theorem gives the $\xi$-BLU predictor for the model under consideration.

**Theorem 5.7.** *Under Model $G_{MR}$, and for known auxiliary measurement vectors, $\mathbf{x}_k' = (x_{k1}, \ldots, x_{kq})$ $(k = 1, \ldots, N)$, the $\xi$-BLU predictor of $\bar{Y}$ is, for any design $p$, given by*

$$T_{BLU} = f_s \bar{Y}_s + (1 - f_s) \mathbf{m}_{\bar{s}}' \hat{\boldsymbol{\beta}}_{BLU} \tag{6.2}$$

*where $\mathbf{m}_{\bar{s}}' = (m_{\bar{s}1}, \ldots, m_{\bar{s}q})$, and for $i = 1, \ldots, q$, $m_{\bar{s}i} = \Sigma_{\bar{s}} x_{ki} / \{N - \nu(s)\}$. Moreover,*

$$\hat{\boldsymbol{\beta}}_{BLU} = (\mathbf{X}_s' \boldsymbol{\Sigma}_s^{-1} \mathbf{X}_s)^{-1} (\mathbf{X}_s' \boldsymbol{\Sigma}_s^{-1} \mathbf{Y}_s)$$

*$\mathscr{E}$MSE$(p, T_{BLU})$ is equal to the p-expectation of*

$$\mathscr{E}(T_{BLU} - \bar{Y})^2 = \frac{\{\Sigma_{\bar{s}} v_k\} \sigma^2}{N^2} + (1 - f_s)^2 \{\mathbf{m}_{\bar{s}}' (\mathbf{X}_s' \boldsymbol{\Sigma}_s^{-1} \mathbf{X}_s)^{-1} \mathbf{m}_{\bar{s}}\} \sigma^2$$

**Proof.** The statement follows by straightforward application of Theorem 5.3: The desired predictor must be of the form $f_s \bar{Y}_s + (1 - f_s) U$, where $\mathscr{E}(U) = \bar{\mu}_{\bar{s}} = \mathbf{m}_{\bar{s}}' \boldsymbol{\beta}$. The $\xi$-unbiased estimator $\hat{\boldsymbol{\beta}}_{BLU}$ of $\boldsymbol{\beta}$ is the well-known generalized least squares estimator; see, for example, Johnston (1972). The expression for $\mathscr{E}(T_{BLU} - \bar{Y})^2$ follows easily. □

**Example 1.** Under the model $\mathscr{E}(Y_k) = \beta$, $\mathscr{V}(Y_k) = \sigma^2$ $(k = 1, \ldots, N)$, $\mathscr{C}(Y_k, Y_l) = 0$ $(k \neq l)$, we have, in the formulation (6.1), $q = 1$, and $x_{k1} = v_k = 1$ for all $k$. Theorem 5.7 yields $T_{BLU} = \bar{Y}_s$ as the $\xi$-*BLU* predictor of $\bar{Y}$, confirming the result of Remark 1, case (iii) of Section 5.2. ■

**Example 2.** Consider the model $\mathscr{E}(Y_k) = \beta_1 + \beta_2 x_k$, $\mathscr{V}(Y_k) = \sigma^2$ $(k = 1, \ldots, N)$, $\mathscr{C}(Y_k, Y_l) = 0$ $(k \neq l)$, that is, the usual linear homoscedastic regression model with one auxiliary variable.

In this case, $q = 2$, $x_{k1} = 1$, $x_{k2} = x_k$, $v_k = 1$ in the formulation (6.1). Theorem 5.7 gives $\hat{\boldsymbol{\beta}}'_{BLU} = (\hat{\beta}_1, \hat{\beta}_2)$, where $\hat{\beta}_1 = \bar{Y}_s - \hat{\beta}_2 \bar{x}_s$, and

$$\hat{\beta}_2 = \frac{\Sigma_s (x_k - \bar{x}_s) Y_k}{\Sigma_s (x_k - \bar{x}_s)^2}$$

Hence, the $\xi$-BLU predictor (6.2) becomes

$$T_{BLU} = f_s \bar{Y}_s + \frac{\{N - \nu(s)\}\hat{\beta}_1 + (\Sigma_{\bar{s}} x_k)\hat{\beta}_2}{N}$$

After simplification we obtain $T_{BLU} = T_{REG}$, where

$$T_{REG} = \bar{Y}_s + \hat{\beta}_2 (\bar{x} - \bar{x}_s)$$

Known in classical survey sampling as (one possible form of) the regression estimator, $T_{REG}$ will be further discussed in Section 7.2. Under the model formulation of this example, $T_{REG}$ has, for *any* design, the $\xi$-BLU property. It is easily shown that

$$\mathscr{E}E(T_{REG} - \bar{Y})^2 = E\left\{\frac{1}{\nu(S)} - \frac{1}{N} + \frac{(\bar{x} - \bar{x}_S)^2}{\Sigma_S (x_k - \bar{x}_S)^2}\right\}\sigma^2$$

By model-based theory, the optimal FES($n$) design would again be purposive, namely, the one that selects with probability one the sample $s$ consisting of those $n$ units whose $x_k$-values minimize the quantity

$$\frac{(\bar{x} - \bar{x}_s)^2}{\Sigma_s (x_k - \bar{x}_s)^2}$$

By contrast, classical sampling theory would recommend that $T_{REG}$ be used in conjunction with simple random sampling or stratified random sampling. The estimator then has a certain, often small $p$-bias; see Cochran (1963, p. 198). ■

We conclude this section by observing that the method of maximum likelihood may be applied, in a fruitful way, to produce solutions to the type of inference problems studied in this section.

The maximum likelihood method was essentially a failure in the Fixed population approach, see Section 2.4. The reason was that the probability structure which generated the likelihood function was too simple to produce anything but trivial conclusions: Any parameter $\mathbf{y} = (y_1, \ldots, y_N)$ consistent with the sample was found to be supported to exactly the same degree by the likelihood function. Nothing is, of course, wrong with a simple structure; in effect, the simplicity would be seen by many as an appealing feature of the Fixed population approach.

Royall (1975) considers the following likelihood approach: Assume that $\mathbf{Y} = \begin{pmatrix} \mathbf{Y}_s \\ \mathbf{Y}_{\bar{s}} \end{pmatrix}$ has an $N$-dimensional normal probability distribution, with mean vector and variance–covariance matrix determined by (6.1). Having observed $\mathbf{Y}_s = \mathbf{y}_s$, the joint likelihood function for $N\bar{y} = \Sigma_1^N y_k$ and the unknown $\beta_1, \ldots, \beta_p$ can be stated without difficulty. The marginal likelihood function (see, for example, Cox and Hinkley, 1974) of $N\bar{y}$, having observed $\mathbf{y}_s$, may then be worked out and maximized to obtain the maximum likelihood predictor of $\bar{Y}$.

Royall (1975) shows that in general the maximum likelihood predictor of $\bar{Y}$, $T_{ML}$, will differ from the $\xi$-BLU predictor $T_{BLU}$ of Theorem 5.7.

In some important special cases, we do, however, have the identity $T_{ML} \equiv T_{BLU}$. A sufficient condition for such identity is that

$$\mathscr{V}(Y_k) = \sigma^2 v_k = \sigma^2 \sum_{i=1}^{q} \gamma_i x_{ki} \qquad (k = 1, \ldots, N)$$

for some $p$-vector $\boldsymbol{\gamma} = (\gamma_1, \ldots, \gamma_p)$. For example, the condition is satisfied if the variance structure is such that

$$\mathscr{V}(Y_k) = \sigma^2 v_k = \sigma^2 x_{ki} \qquad (k = 1, \ldots, N) \tag{6.3}$$

for *some* variable $x_i$ $(i = 1, \ldots, q)$.

**Example 3**. The condition (6.3) is satisfied for the two models dealt with in Examples 1 and 2, so that in fact $T_{ML} = T_{BLU}$ in those two cases.

However, as an example of nonidentity, consider the model $\mathscr{E}(Y_k) = \beta x_k$, $\mathscr{V}(Y_k) = \sigma^2$ $(k = 1, \ldots, N)$, $\mathscr{C}(Y_k, Y_l) = 0$ $(k \neq l)$. The condition (6.3) is not met (unless all $x_k$ are equal).

The $\xi$-BLU predictor is given by (4.1) with $v(x) = 1$ in (4.2). By contrast, Royall's maximum likelihood predictor for that model turns out to be

$$T_{ML} = \bar{Y}_s + \hat{\beta}_2(\bar{x} - \bar{x}_s)$$

which is identical to the $\xi$-BLU predictor in Example 2 above. ■

## 5.7. A SUMMARY OF RESULTS ON THE SAMPLE MEAN

Because of its simplicity and intuitive appeal, the sample mean predictor $\bar{Y}_s$ merits some special attention. We have already observed that $\bar{Y}_s$ emerges as optimal under several model formulations and under various criteria. Our objective now is to add to and to summarize the picture with respect to $\bar{Y}_s$. A review of results on $\bar{Y}_s$ is given in table form at the end of the section.

Recall first that in Chapter 4 (see Table 4.2) we showed that $\bar{Y}_s$ forms an optimal *p-unbiased strategy* when combined with an FES($n$) design having the inclusion probabilities $\alpha_k = f\ (k = 1, \ldots, N)$. These results obtained under Model $G_{T0}$ (for linear predictors) and Model $E_{T0}$ (without restriction to linearity), that is, in situations where the labels are uninformative in the sense that the distribution of $Y_k$ is characterized by the model in exactly the same way for every label $k$.

In Chapter 5, the results of Remark 1, Section 5.2 and Theorem 5.4 deal with the optimality of $\bar{Y}_s$ in classes of *ξ-unbiased predictors*. This optimality holds for *any* design under Model $G_{T0}$ (when the predictors are required to be linear) and under Model $E_{PI}$ (when there is no linearity restriction). Again, these are cases where the distributional properties of the $Y_k$ are identical for all labels $k$.

The concept of a $p\xi$-unbiased predictor of $\bar{Y}$ was introduced in Section 4.3, but so far we have not pursued the problem of finding minimum $\mathscr{E}$ MSE predictors in the class of $p\xi$-unbiased predictors, with or without restriction to linearity. It was pointed out that any $p$-unbiased predictor is automatically $p\xi$-unbiased, and so is any $\xi$-unbiased predictor.

The wider class of $p\xi$-unbiased predictors provides an interesting object of study. Here we shall limit ourselves to showing, for an FES($n$) design, that $\bar{Y}_s$ retains its optimality properties under Model $G_{T0}$ (for linear predictors) and under Model $E_{PI}$ (without linearity restriction) when the class of predictors is expanded by relaxing the requirements to $p\xi$-unbiasedness. The details are spelled out in the following theorem.

We admit as a predictor of $\bar{Y}$ any function $T = T(\mathscr{D})$ of the random variable $\mathscr{D} = \{(k, Y_k): k \in S\}$, which is subject to the combined randomness of $S$ and of the $Y_k$. If $T$ is represented on the form (1.3), it is easy to see that, for any FES($n$) design $p$, $T$ is a $p\xi$-unbiased predictor of $\bar{Y}$ if and only if $U = U(\mathscr{D})$ is a $p\xi$-unbiased predictor of $\bar{Y}_{\bar{s}}$.

**Theorem 5.8.** *Let p be any given FES(n) design with* $\alpha_k > 0\ (k = 1, \ldots, N)$. *Then the following holds:*

*(i) Under Model* $E_{PI}$,

$$\mathscr{E}E(T - \bar{Y})^2 \geqslant \mathscr{E}E(\bar{Y}_S - \bar{Y})^2 = \frac{(1-f)\sigma^2}{n}$$

*for any pξ-unbiased predictor T; equality holds if and only if* $T = \bar{Y}_s$;

*(ii) Under Model* $G_{T0}$,

$$\mathscr{E}E(T - \bar{Y})^2 \geqslant \mathscr{E}E(\bar{Y}_S - \bar{Y})^2 = \frac{(1-\rho)(1-f)\sigma^2}{n}$$

*for any linear pξ-unbiased predictor T; equality holds if and only if* $T = \bar{Y}_s$.

**Proof.** In case (i) let $T = f\bar{Y}_s + (1-f)U$. The condition $\mathscr{E}E(T) = \mu$ implies $\mathscr{E}E(U) = \mu$, where $\mu = \mathscr{E}(Y_k)\ (k = 1, \ldots, N)$. Formula (1.6), in which the covariance term vanishes, can be written as

$$\mathscr{E}E(T - \bar{Y})^2 = (1-f)^2 \left\{ \mathscr{E}E(U-\mu)^2 + \frac{\sigma^2}{N-n} \right\}$$

We must find $U = U(\mathscr{D})$ to minimize the $p\xi$-variance $\mathscr{E}E(U-\mu)^2$ subject to $\mathscr{E}E(U) = \mu$, where $\mathscr{D} = \{(k, Y_k) : k \in S\}$. But Model $\mathrm{E}_{PI}$ is a special case of Model $\mathrm{E}_T$ such that $a_k = 1$, $b_k = 0$ $(k = 1, \ldots, N)$ and the $Y_k$ are independent, so that $\rho = 0$. Hence, applying Lemma 4.5 to this special case (which makes $T_{GDo} = \bar{Y}_s$), we conclude that $U = \bar{Y}_s$ is the minimum $p\xi$-variance $p\xi$-unbiased predictor of $\mu$, and $\mathscr{E}E(\bar{Y}_S - \mu)^2 = \sigma^2/n$. Therefore $T = f\bar{Y}_s + (1-f)\bar{Y}_s = \bar{Y}_s$ is the optimal predictor in case (i).

In case (ii), consider formula (3.2) of Section 4.3 which may be written

$$\mathscr{E}E(T - \bar{Y})^2 = \mathscr{E}E(T-\mu)^2 + \mathscr{V}(\bar{Y}) - 2\mathscr{E}E(T-\mu)(\bar{Y}-\mu)$$

Now, $\mathscr{E}E(T) = \mu$ and $T = w_{0s} + \Sigma_s w_{ks} Y_k$ imply $E(w_{0S}) = 0$ and $E(\Sigma_S w_{kS}) = 1$. Use of the latter equation gives $\mathscr{E}E(T-\mu)(\bar{Y}-\mu) = \{1 + (N-1)\rho\}\sigma^2/N$, which is independent of $T$ and equal to the value of $\mathscr{V}(Y)$. Moreover, $\mathscr{E}E(T-\mu)^2$ is, by Lemma 4.2, minimized by $T = \bar{Y}_s$, which completes the proof. □

**Example 1.** Consider the $p$-unbiased Horvitz–Thompson predictor $T_{HT}$, formula (4.2) of Section (4.4). Now, $T_{HT}$ is also $p\xi$-unbiased under any model. In particular, under Model $\mathrm{E}_{PI}$ and for any given FES($n$) design with $\alpha_k > 0$ $(k = 1, \ldots, N)$, we have from Theorem 5.8.

$$\mathscr{E}E(T_{HT} - \bar{Y})^2 \geqslant \frac{(1-f)\sigma^2}{n}$$

**Table 5.1. Summary of Optimality Results Involving $\bar{Y}_s$ under the $\mathscr{E}$MSE Criterion.**

| Unbiasedness Condition | Model | Conditions on Optimal Design | Class in which Optimality Obtains |
|---|---|---|---|
| $p$-unbiased | $\mathrm{G}_{T0}$ | FES($n$) with $\alpha_k = f$, all $k$ | All $p$-unbiased strategies $(p, T)$ where $T$ is linear |
| $p$-unbiased | $\mathrm{E}_{T0}$ | FES($n$) with $\alpha_k = f$, all $k$ | All $p$-unbiased strategies $(p, T)$ |
| $\xi$-unbiased | $\mathrm{G}_{T0}$ | None | All linear $\xi$-unbiased $T$ |
| $\xi$-unbiased | $\mathrm{E}_{PI}$ | None | All $\xi$-unbiased $T$ |
| $p\xi$-unbiased | $\mathrm{G}_{T0}$ | FES($n$) | All linear $p\xi$-unbiased $T$ |
| $p\xi$-unbiased | $\mathrm{E}_{PI}$ | FES($n$) | All $p\xi$-unbiased $T$ |

with equality if and only if the design has the inclusion probabilities $\alpha_k = f\ (k = 1, \ldots, N)$, in which case $T_{HT} = \bar{Y}_s$. ■

**Remark 1.** Case (i) of Theorem 5.8 was shown by Wretman (1970). Various versions of case (ii) were obtained by Scott and Smith (1969a) (assuming the $Y_k$ to be uncorrelated), J. N. K. Rao (1975) (the case of $\rho = -1/(N-1)$), and Rao and Bellhouse (1976). The latter paper contains additional results on linear $p\xi$-unbiased predictors, for example, for stratified sampling designs.

Further results on the optimality of $\bar{Y}_s$ under various conditions include the following: Kempthorne (1969) showed that $\bar{Y}_s$ has minimum $\mathscr{E}$MSE under Model $\mathrm{E}_{R\,\rho 0}$ in the class of linear $p$-unbiased and translation invariant predictors. C. R. Rao (1971) showed that the condition of translation invariance is unnecessary. Finally, Ramakrishnan (1970) required translation invariance but relaxed the $p$-unbiasedness assumption. ■

Table 5.1 gives a summary of results on $\bar{Y}_s$. Recall also that $\bar{Y}_s$ is minimax estimator under certain conditions; see Section 3.6. Moreover, $\bar{Y}_s$ will emerge as the best choice of predictor under the robustness considerations to be considered in Section 7.7.

CHAPTER 6

# Inference under Superpopulation Models: Using Tools of Bayesian Inference

This chapter is similar to the preceding one in that inference is justified by a superpopulation assumption. In Bayesian terminology we can say that the superpopulation appears in the form of a prior distribution for the components of $\mathbf{y} = (y_1, \ldots, y_N)$, considered as a fixed unknown parameter vector. Given such a prior, and given the data $d = \{(k, y_k) : k \in s\}$, we can then derive the posterior of $y_1, \ldots, y_N$, and make a standard Bayesian inference about the unknown population mean $\bar{y} = \Sigma_1^N y_k/N$.

## 6.1. BASIC THEORY IN THE BAYESIAN APPROACH

Following Ericson (1969a), we shall develop the Bayesian approach under noninformative designs, as defined in Section 1.4. This results in an inference that is independent of the design $p$. Hence in this "purely Bayesian case" (Solomon and Zacks, 1970), randomized sampling designs play no role, and the sample could have been chosen in a purposive way, but a randomly selected sample is probably a sounder base for the inference, as argued in Chapter 5.

Bayesian procedures in which the optimal selection procedure is sequential and informative have been considered by Basu (1969), Zacks (1969), Solomon and Zacks (1970). In such sampling designs, the selection probabilities in a given draw may depend on the $y$-values of already drawn units. Zacks (1969) studied Bayes sequential designs for a fixed sample size $n$, and derived minimum posterior risk designs for certain situations. The results indicate that it is difficult to implement the optimal sequential designs. We shall only discuss noninformative designs.

Let $\mathbf{y} = (y_1, \ldots, y_N)$ be an unknown but fixed parameter. Let $g(\mathbf{y})$ denote the prior of $y_1, \ldots, y_N$, that is, $g(\mathbf{y})$ is an assessment, obtained by the methods usually used by Bayesians, of the statistician's subjective prior belief about $\mathbf{y}$. Let $\bar{s} = \{1, \ldots, N\} - s$ be the set of units not in the sample $s$, let $g_s(\mathbf{y}_s)$ be the marginal density of $g(\cdot)$ for the units $k \in s$, enumerated in order of increasing $k$, and let $g_{\bar{s}|s}(\mathbf{y}_{\bar{s}}) = g(\mathbf{y})/g_s(\mathbf{y}_s)$ be the conditional density for the units $k \in \bar{s}$, in order of increasing $k$, given $d = \{(k, y_k) : k \in s\}$. Note that if $g(\cdot)$ is an exchangeable density, then $g_s(\cdot)$ represents one and the same symmetric function for all $s$ of a given fixed effective size, and so is the case with $g_{\bar{s}\,|s}(\cdot)$.

Assuming $p$ to be noninformative, the Bayesian argument is developed as follows:

The sampling design defines the likelihood element in obtaining the data $d = \{(k, y_k) : k \in s\}$, that is, using the concept of likelihood as in Section 2.4,

$$L_d(\mathbf{y}) = \begin{cases} p(s) & \text{for } \mathbf{y} \in \Omega_d \\ 0 & \text{otherwise} \end{cases}$$

where $\Omega_d$ is the subspace of $\Omega$ consistent with $d$. Employing $g(\mathbf{y})$ as a prior for $\mathbf{y}$, the posterior is

$$g(\mathbf{y} \mid d) \propto \begin{cases} p(s)\, g(\mathbf{y}) & \text{for } \mathbf{y} \in \Omega_d \\ 0 & \text{otherwise} \end{cases} \tag{1.1}$$

Since $p$ is noninformative, the deleted factor of proportionality in (1.1) is unity divided by

$$\int p(s)\, g(\mathbf{y})\, d\mathbf{y}_{\bar{s}} = p(s)\, g_s(\mathbf{y}_s)$$

where $\int$ is over $\{\mathbf{y}_{\bar{s}} : -\infty < y_k < \infty,\ k \in \bar{s}\}$. Hence, thanks to the noninformativity of $p$, the simple result is

$$g(\mathbf{y} \mid d) = \begin{cases} g(\mathbf{y})/g_s(\mathbf{y}_s) & \text{for } \mathbf{y} \in \Omega_d \\ 0 & \text{otherwise} \end{cases} \tag{1.2}$$

This posterior distribution of $y_1, \ldots, y_N$, given $d$, is the key to the Bayesian inference.

Now, (1.2) is a degenerate $N$-dimensional density with all the probability mass concentrated in the subspace $\Omega_d$. Clearly, $g(\mathbf{y})/g_s(\mathbf{y}_s) = g_{\bar{s}|s}(\mathbf{y}_{\bar{s}})$ represents the posterior opinion about the unobserved coordinates of $\mathbf{y}$. In general, $g_{\bar{s}|s}(\mathbf{y}_{\bar{s}})$ depends on $d = \{(k, y_k) : k \in s\}$, but under certain assumptions on $g(\cdot)$, the Bayesian inference obtained by means of (1.2) will depend only on the unlabeled data, $y_s = \{y_k : k \in s\}$, as we shall see.

Since $p$ does not enter into (1.2), the design will play no role in making Bayesian inference about any function of $y_1, \ldots, y_N$. The structure of $g(\cdot)$ alone will determine the inference. Likewise, in Chapter 5, noninformativity of the design gave classical inferences, in Theorems 5.1 and 5.3, which depended only on the superpopulation and not on the design.

The prior that we assume for $y_1, \ldots, y_N$ is essential for determining the Bayesian inference. We consider two different bases for choice of prior, Case A and Case B below. The Bayesian inferences arising from the two cases will be analyzed in the subsequent sections.

*Case A.* Let $g(\mathbf{y} \mid \boldsymbol{\theta})$ be a density function indexed by a known parameter $\boldsymbol{\theta}$, and take $g(\mathbf{y}) = g(\mathbf{y} \mid \boldsymbol{\theta})$ as the prior density of $y_1, \ldots, y_N$ in (1.2).

In general, $g(\cdot \mid \boldsymbol{\theta})$ may be any density, but particularly interesting cases include the ones where $g(\cdot \mid \boldsymbol{\theta})$ is specified in some general terms, as in the cases where $g(\cdot \mid \boldsymbol{\theta})$ has the properties of the superpopulation models $\mathrm{G}_{PI}$, $\mathrm{E}_P$, and $\mathrm{E}_{PI}$ of Section 4.2, that is, $g(\cdot \mid \boldsymbol{\theta})$ is exchangeable and/or composed as a product density. More detailed specification would, in addition, incorporate an assumption of the shape of $g(\cdot \mid \boldsymbol{\theta})$.

In summary, Case A can be described as one where a superpopulation density $g(\mathbf{y} \mid \boldsymbol{\theta})$ with known indexing parameter $\boldsymbol{\theta}$ is playing the role of a prior for $y_1, \ldots, y_N$; see Section 6.2.

*Case B.* Let $\boldsymbol{\theta}$ be unknown in the otherwise known density $g(\cdot \mid \boldsymbol{\theta})$. Following conventional methods of assessing priors, let us assign $\boldsymbol{\theta}$ the known prior distribution $F(\boldsymbol{\theta} \mid \boldsymbol{\phi})$, with a specified value of the parameter $\boldsymbol{\phi}$. The mixture distribution (2.4) of Section 4.2 is taken as the prior of $y_1, \ldots, y_N$, that is, in (1.2), set $g(\mathbf{y}) = {}_1g(\mathbf{y} \mid \boldsymbol{\phi})$, where

$$
{}_1g(\mathbf{y} \mid \boldsymbol{\phi}) = \int_{\Theta} g(\mathbf{y} \mid \boldsymbol{\theta})\, dF(\boldsymbol{\theta} \mid \boldsymbol{\phi}) \tag{1.3}
$$

Again, cases of particular interest include those where $g(\cdot \mid \boldsymbol{\theta})$ is exchangeable and/or a product density, as in Models $\mathrm{G}_{PI}$, $\mathrm{E}_P$, and $\mathrm{E}_{PI}$.

Note that in Case B there will be, following Ericson (1969a), a double usage of the terms "prior" and "posterior": ${}_1g(\mathbf{y} \mid \boldsymbol{\phi})$ and ${}_1g(\mathbf{y} \mid d, \boldsymbol{\phi})$ will be called the prior and the posterior, respectively, of the unknown parameter vector $\mathbf{y}$. In addition, we shall be concerned with the prior $F(\boldsymbol{\theta} \mid \boldsymbol{\phi})$ and the posterior $F(\boldsymbol{\theta} \mid d, \boldsymbol{\phi})$ of the unknown parameter vector $\boldsymbol{\theta}$.

In summary, Case B can be described as one where $g(\mathbf{y} \mid \boldsymbol{\theta})$ is a superpopulation density, conditional on an unknown $\boldsymbol{\theta}$, and where $\boldsymbol{\theta}$ is assigned a prior distribution, $F(\boldsymbol{\theta} \mid \boldsymbol{\phi})$; the mixing distribution (1.3) forms a new superpopulation density which plays the role of a prior for $y_1, \ldots, y_N$; see Section 6.3.

## 6.2. BAYESIAN INFERENCE WHEN $\boldsymbol{\theta}$ IS KNOWN

Let $g(\mathbf{y} \mid \boldsymbol{\theta})$ be the prior density of $y_1, \ldots, y_N$; $\boldsymbol{\theta}$ is assumed known. For a given $s \in \mathscr{S}$, $g_s(\mathbf{y}_s \mid \boldsymbol{\theta})$ will denote the marginal density of $y_k$ for $k \in s$, $k$ in increasing order, while $g_{\bar{s}|s}(\mathbf{y}_{\bar{s}} \mid \boldsymbol{\theta})$ will denote the conditional density of $y_k$ for $k \in \bar{s}$, $k$ in increasing order, given $d$. Also, $\mathscr{E}$, $\mathscr{E}_s$, and $\mathscr{E}_{\bar{s}|s}$ will denote the expectation operators with respect to these three densities, respectively.

As is well known, the Bayesian inference under quadratic loss consists simply in deriving the posterior expectation of $\bar{y}$, given $d$ and $\boldsymbol{\theta}$. Hence we arrive at the *Bayes estimate*; see for example Godambe (1969a),

$$\mathscr{E}(\bar{y} \mid d, \boldsymbol{\theta}) = \frac{\Sigma_s y_k + \psi(d, \boldsymbol{\theta})}{N} \tag{2.1}$$

where $\psi(d, \boldsymbol{\theta}) = \mathscr{E}_{\bar{s}|s}(\Sigma_{\bar{s}} Y_k)$, and $d = \{(k, y_k): k \in s\}$ .

The same inference is obtained by considering the problem as one of prediction of $\bar{Y}$, the objective being to minimize the $\mathscr{E}$MSE criterion (1.1) of Section 5.1. If $T = T(d)$ is a predictor of $\bar{Y}$ and $Y_1, \ldots, Y_N$ have the specified superpopulation distribution $g(\mathbf{y} \mid \boldsymbol{\theta})$, we know from Section 5.1, formulas (1.3) and (1.5), that the minimum $\mathscr{E}$MSE predictor is given by

$$T^* = \frac{\Sigma_s Y_k + \psi(d, \boldsymbol{\theta})}{N} \tag{2.2}$$

where $d = \{(k, Y_k) : k \in s\}$ and the function $\psi(\cdot, \boldsymbol{\theta})$ is as in (2.1). The only difference between (2.1) and (2.2) is one of interpretation: $T^*$ is seen as a predictor, that is, a function of the random variables $Y_k$ for $k \in s$, while the Bayes estimate (2.1) is the value taken by the predictor $T^*$ for the outcome $d = d$, that is, when $Y_k = y_k, k \in s$.

We summarize these findings in the following theorem.

**Theorem 6.1.** *Let $p$ be any given design. Let $T$ be any predictor of $\bar{Y}$. Then*

$$\mathscr{E}\text{MSE}(p, T) \geqslant \mathscr{E}\text{MSE}(p, T^*)$$

*where, for any $s$, $T^*$ is given by (2.2). The value (2.1) taken by $T^*$ for the outcome $d = d$ is the Bayes estimate of $\bar{y}$.* □

**Remark 1.** Concerning the Bayesian inference (2.1), note: (a) that it is independent of the design, (b) that the uncertainty of (2.1) may be expressed in terms of the variance of its posterior distribution, (c) that $T^*$ given by (2.2) is $\xi$-unbiased for $\bar{Y}$: For any $s$,

$$\mathscr{E}(T^*) = \frac{\Sigma_s \mu_k + \mathscr{E}_s\, \mathscr{E}_{\bar{s}|s}(\Sigma_{\bar{s}} Y_k)}{N} = \bar{\mu}$$

However, this latter property is of limited interest for the Bayesian statistician. ■

We analyze the nature of the inference (2.1) in three situations, namely, when $g(\cdot \mid \boldsymbol{\theta})$ satisfies the requirements of Model $\text{G}_{PI}$, $\text{E}_P$, or $\text{E}_{PI}$. Even if the functional form of $g(\cdot \mid \boldsymbol{\theta})$ is unknown, these models still express certain crude prior knowledge.

*Model $G_{PI}$.* $g(\cdot \mid \boldsymbol{\theta})$ is a product density, not necessarily exchangeable. Because of the independence of the $Y_k$, the conditional density $g_{\bar{s}|s}(\cdot \mid \boldsymbol{\theta})$ equals the marginal density $g_{\bar{s}}(\cdot \mid \boldsymbol{\theta})$. Letting $\mathscr{E}(Y_k) = \mu_k = \mu_k(\boldsymbol{\theta})$, we have

$$\psi(d, \boldsymbol{\theta}) = \Sigma_{\bar{s}} \mu_k(\boldsymbol{\theta})$$

Thus beyond its dependence on $\boldsymbol{\theta}$, the Bayes estimate (2.1) depends on $d$ only through the pair $(s, \Sigma_s y_k)$.

*Model $E_P$.* $g(\cdot \mid \boldsymbol{\theta})$ is exchangeable. In this case, the function $g_{\bar{s}|s}(\cdot \mid \boldsymbol{\theta})$ is symmetric. Hence, the function $\psi(d, \boldsymbol{\theta})$ in (2.1) is symmetric in the $y_k$ for $k \in s$. In other words, the Bayes estimate (2.1) depends on $\boldsymbol{\theta}$ and on $d$ only through the unlabeled data $y_s = \{y_k : k \in s\}$. But it cannot be claimed that we can further reduce $y_s$ and still make the optimal Bayesian inference under this model.

*Model $E_{PI}$.* $g(\cdot \mid \boldsymbol{\theta})$ is an exchangeable product density. In this case, $\mathscr{E}(Y_k) = \mu = \mu(\boldsymbol{\theta})$ for all $k$. Thus,

$$\psi(d, \boldsymbol{\theta}) = \{N - \nu(s)\} \mu(\boldsymbol{\theta})$$

Hence, the Bayes estimate (2.1) depends on $\boldsymbol{\theta}$ and on $d$ through the pair $(\nu(s), \Sigma_s y_k)$ only, which is a further reduction of the unlabeled data $y_s$.

**Remark 2.** In the cases above, the Bayesian inference depends on the data $d = \{(k, y_k) : k \in s\}$ only through certain reductions of $d$. The reduction of $d$ obtained in a given case is, therefore, in a sense "sufficient"; the inference can be obtained from knowledge of the reduction alone. This can be formalized through the concept of Bayes sufficiency, Godambe (1968), Godambe and Thompson (1971a), Solomon and Zacks (1970).

Assume as in Section 2.1 that the data $d$, an outcome of $D$, are realized with probability $p(s)$ if $\mathbf{y} \in \Omega_d$ and with probability zero otherwise, where $\Omega_d$ is the set of all $\mathbf{y} \in R_N$ such that $d$ is consistent with $\mathbf{y}$. Let $t(d)$ be a statistic on $\{d : s \in \mathscr{S}, \mathbf{y} \in R_N\}$. Then $t(d)$ is defined to be *Bayes sufficient* for $\bar{y}$ if and only if, for every member $\xi$ of a given class of priors of $\mathbf{y}$, the posterior of $\bar{y}$, given $d$, depends on $d$ only through $t(d)$. Thus $t(d)$ can be said to contain all of the statistician's posterior knowledge about $\bar{y}$. It turns out that for Models $G_{PI}$, $E_P$, and $E_{PI}$, the statistics $(s, \Sigma_s y_k), y_s$, and $(\nu(s), \Sigma_s y_k)$, respectively, are Bayes sufficient. In particular, the posterior mean of $\bar{y}$, given $d$, must then also be a function of these reductions of $d$, as we have seen above.

Godambe (1968) used the Bayes sufficiency of $(s, \Sigma_s y_k)$ obtained from Model $G_{PI}$ as a justification for limiting consideration to estimators that depend on $d$ only in this condensed fashion. This enabled him to show some strong optimality properties: The sample mean $\bar{y}_s$ is the uniquely Bayes sufficient, origin and scale invariant (see Section 3.7) estimator of the population mean $\bar{y}$. Moreover, the difference estimator $t_D$ and the ratio estimator $t_R$ depend on $d$ through $(s, \Sigma_s y_k)$ only, as pointed out in Remark 1 of Section 1.6. The former estimator is origin invariant and the latter is scale invariant. Combining the Bayes sufficiency under Model $G_{PI}$ with the respective invariance properties, Godambe (1968) obtained the optimality properties of $t_D$ and $t_R$ stated in Remark 1 of Section 3.7.

Thus we see that traditional estimators such as $t_D$ and $t_R$ can be justified by a combination of criteria (Bayes sufficiency and invariance), which permits a unique estimator to be singled out. It is true, however, that these results depend heavily on the assumed independence of the $Y_k$ postulated by Model $G_{PI}$, as pointed out by J. N. K. Rao (1971). ■

## 6.3. BAYESIAN INFERENCE WHEN $\boldsymbol{\theta}$ IS UNKNOWN

This is Case B of Section 6.1; it may be seen as a Bayesian treatment of the inference problem posed and solved in Chapter 5 through classical inference tools. The Bayesian approach will bring in an additional probabilistic layer in the form of the prior $F(\boldsymbol{\theta} \mid \boldsymbol{\phi})$ on the parameter space $\Theta$.

Take ${}_1g(\mathbf{y} \mid \boldsymbol{\phi})$ defined by (1.3) to be the prior for $y_1, \ldots, y_N$, where $g(\cdot \mid \boldsymbol{\theta})$ as well as $F(\cdot \mid \boldsymbol{\phi})$ are known functions and $\boldsymbol{\phi}$ is a specified parameter vector.

For given data $d = \{(k, y_k) : k \in s\}$, let ${}_1g_s(\mathbf{y}_s \mid \boldsymbol{\phi})$ be the $\nu(s)$-dimensional marginal prior density of ${}_1g(\cdot \mid \boldsymbol{\phi})$ of $y_k$ for $k \in s$, $k$ taken in increasing order of magnitude.

Let ${}_1g_{\bar{s}|s}(\mathbf{y}_{\bar{s}} \mid \boldsymbol{\phi})$ be the $\{N - \nu(s)\}$ – dimensional conditional prior density, given $d$, of $y_k$, $k \in \bar{s}$, $k$ taken in increasing order. In other words,

$$ {}_1g_{\bar{s}|s}(\mathbf{y}_{\bar{s}} \mid \boldsymbol{\phi}) = \frac{{}_1g(\mathbf{y} \mid \boldsymbol{\phi})}{{}_1g_s(\mathbf{y}_s \mid \boldsymbol{\phi})} $$

Let ${}_1\mathscr{E}$, ${}_1\mathscr{E}_s$, and ${}_1\mathscr{E}_{\bar{s}|s}$ denote, respectively, the expectation operators with respect to ${}_1g(\cdot \mid \boldsymbol{\phi})$, ${}_1g_s(\cdot \mid \boldsymbol{\phi})$, and ${}_1g_{\bar{s}|s}(\cdot \mid \boldsymbol{\phi})$.

The Bayesian inference in this situation is like that of Theorem 6.1 with one small but important modification: The expectation operators $\mathscr{E}$, $\mathscr{E}_s$, and $\mathscr{E}_{\bar{s}|s}$, which are conditional on $\boldsymbol{\theta}$, must be replaced by ${}_1\mathscr{E}$, ${}_1\mathscr{E}_s$, and ${}_1\mathscr{E}_{\bar{s}|s}$, respectively. Moreover, we must replace $\psi(d, \boldsymbol{\theta})$ by

$$ {}_1\psi(d, \boldsymbol{\phi}) = {}_1\mathscr{E}_{\bar{s}|s}(\Sigma_{\bar{s}} Y_k) \tag{3.1} $$

We then obtain the following theorem.

**Theorem 6.2.** *Let $p$ be any given design, and let $T$ be any predictor of $\bar{Y}$. Then*

$$ \mathscr{E}\,\mathrm{MSE}(p, T) \geqslant \mathscr{E}\,\mathrm{MSE}(p, T^*) $$

*where, for any s,*

$$T^* = \frac{\Sigma_s Y_k + {}_1\psi(d,\boldsymbol{\phi})}{N}$$

*where* ${}_1\psi(\cdot, \boldsymbol{\phi})$ *is given by (3.1). The value taken by* $T^*$ *for the outcome* $d = d$,

$$t^* = \frac{\Sigma_s y_k + {}_1\psi(d,\boldsymbol{\phi})}{N} \tag{3.2}$$

*is the Bayes estimate of* $\bar{y}$. □

This theorem may be viewed as the Bayesian counterpart of Theorems 5.1 and 5.3 in the classical approach: It gives the optimal Bayesian inference when the superpopulation distribution, conditional on an unknown $\boldsymbol{\theta}$, is $g(\mathbf{y} \mid \boldsymbol{\theta})$, and $\boldsymbol{\theta}$ is assigned a prior, $F(\boldsymbol{\theta} \mid \boldsymbol{\phi})$, with density function $f(\boldsymbol{\theta} \mid \boldsymbol{\phi})$.

We now examine the nature of the Bayes estimate (3.2) under various assumptions on $g(\cdot \mid \boldsymbol{\theta})$. A different perspective on $t^*$ is gained if we note that the expectation in (3.1) is to be taken with respect to the density

$${}_1g_{\bar{s}|s}(\mathbf{y}_{\bar{s}} \mid \boldsymbol{\phi}) = \frac{{}_1g(\mathbf{y} \mid \boldsymbol{\phi})}{{}_1g_s(\mathbf{y}_s \mid \boldsymbol{\phi})} = \int_{\Theta} g_{\bar{s}|s}(\mathbf{y}_{\bar{s}} \mid \boldsymbol{\theta})\, dF(\boldsymbol{\theta} \mid d, \boldsymbol{\phi})$$

where

$$dF(\boldsymbol{\theta} \mid d, \boldsymbol{\phi}) = \frac{g_s(\mathbf{y}_s \mid \boldsymbol{\theta})\, dF(\boldsymbol{\theta} \mid \boldsymbol{\phi})}{{}_1g_s(\mathbf{y}_s \mid \boldsymbol{\phi})}$$

is the posterior of $\boldsymbol{\theta}$, given $d = \{(k, y_k) : k \in s\}$ and $\boldsymbol{\phi}$. This follows by use of (1.3) and the identity $g(\mathbf{y} \mid \boldsymbol{\theta}) = g_s(\mathbf{y}_s \mid \boldsymbol{\theta}) g_{\bar{s}|s}(\mathbf{y}_{\bar{s}} \mid \boldsymbol{\theta})$.

In particular, consider now Models $E_P$ and $E_{PI}$:

***Model*** $E_P$. $g(\cdot \mid \boldsymbol{\theta})$ is exchangeable. In this case, $F(\boldsymbol{\theta} \mid d, \boldsymbol{\phi})$ depends on $d$ only through the unlabeled data $y_s = \{y_k : k \in s\}$. It follows that ${}_1g_{\bar{s}|s}(\cdot \mid \boldsymbol{\phi})$ is also exchangeable. Therefore, the Bayes estimate (3.2) will depend on $\boldsymbol{\theta}$ and on $d$ through $y_s$ only.

***Model*** $E_{PI}$. $g(\cdot \mid \boldsymbol{\theta})$ is an exchangeable product density. Letting

$$\mu(\boldsymbol{\theta}) = \mathscr{E}(Y_k) = \int_{-\infty}^{\infty} y_k g(\mathbf{y} \mid \boldsymbol{\theta})\, dy_k$$

be the common mean of the $Y_k$, conditional on $\boldsymbol{\theta}$, we get

$${}_1\mathscr{E}_{\bar{s}|s}(\Sigma_{\bar{s}} Y_k) = {}_1\mathscr{E}_{\bar{s}}(\Sigma_{\bar{s}} Y_k) = \{N - \nu(s)\} E_{\boldsymbol{\theta}}\{\mu(\boldsymbol{\theta}) \mid d, \boldsymbol{\phi}\}$$

where

$$E_{\boldsymbol{\theta}}\{\mu(\boldsymbol{\theta}) \mid d, \boldsymbol{\phi}\} = \int \mu(\boldsymbol{\theta})\, dF(\boldsymbol{\theta} \mid d, \boldsymbol{\phi}) \tag{3.3}$$

is the expectation of $\mu(\boldsymbol{\theta})$ with respect to the posterior of $\boldsymbol{\theta}$, $F(\boldsymbol{\theta} \mid d, \boldsymbol{\phi})$. Because of the exchangeability, (3.3) depends on $d$ through the unlabeled data $y_s$ only.

The Bayes estimate (3.2) takes the following form given in Ericson (1969a),

$$t^* = f_s\bar{y}_s + (1 - f_s)E_{\boldsymbol{\theta}}\{\mu(\boldsymbol{\theta}) \mid d, \boldsymbol{\phi}\} \tag{3.4}$$

where $E_{\boldsymbol{\theta}}\{\mu(\boldsymbol{\theta}) \mid d, \boldsymbol{\phi}\}$ is given by (3.3). In the next section we shall work out the estimate (3.4) in several examples involving specified shapes of the functions $g(\cdot \mid \boldsymbol{\theta})$ and $F(\cdot \mid \boldsymbol{\phi})$. We have concluded that (3.4) depends on $d$ through $y_s$ only, but under the normal prior $g(\cdot \mid \boldsymbol{\theta})$ to be assumed, $t^*$ can be obtained by even further reduction of $d$.

## 6.4. EXAMPLES OF BAYESIAN INFERENCE UNDER NORMAL PRIORS

We illustrate the Bayesian approach by means of several examples. The first three are adapted from Ericson (1969a). We use the results of Section 6.3, Model $\text{E}_{PI}$, that is, the prior of $\mathbf{y}$ is a product density, $g(\mathbf{y} \mid \boldsymbol{\theta}) = \Pi_1^N g(y_k \mid \boldsymbol{\theta})$. The Bayes estimate will then be given by (3.4). For $g(y_k \mid \boldsymbol{\theta})$, a normal distribution is assumed. Mathematically expedient choices of $F(\boldsymbol{\theta} \mid \boldsymbol{\phi})$ are considered.

**Example 1.** Let $g(y_k \mid \theta)$ $(k = 1, \ldots, N)$ be $N(\theta, \sigma_a^2)$, where $\theta$ is unknown and $\sigma_a^2$ known. Let $F(\theta \mid \theta_0, \sigma_0^2)$ be $N(\theta_0, \sigma_0^2)$, where $\boldsymbol{\phi} = (\theta_0, \sigma_0^2)$ is known. Under these conditions the exchangeable prior (1.3) is $N$-dimensional normal with common means, $\theta_0$, common variances, $\sigma_a^2 + \sigma_0^2$, and common covariances, $\sigma_0^2$. The implied prior of $\bar{y} = \Sigma_1^N y_k/N$ is $N(\theta_0, \sigma_0^2 + \sigma_a^2/N)$.

Given $d = \{(k, y_k): k \in s\}$, the posterior of $y_k$, $k \in \bar{s}$, ${}_1g_{\bar{s}|s}(\cdot)$, is $\{N - \nu(s)\}$-dimensional normal with common means,

$$\theta_1 = \frac{\sigma_a^2\theta_0 + \nu(s)\sigma_0^2\bar{y}_s}{\sigma_a^2 + \nu(s)\sigma_0^2}$$

common variances, $\sigma_a^2 + \sigma_1^2$, and common covariances, $\sigma_1^2$, where

$$\sigma_1^2 = 1\bigg/\left\{\frac{\nu(s)}{\sigma_a^2} + \frac{1}{\sigma_0^2}\right\}$$

Hence, ${}_1\mathscr{E}_{\bar{s}|s}(\bar{y}_{\bar{s}}) = \theta_1$, and using (3.4), the Bayes estimate of $\bar{y}$ is

$$t_1^* = f_s\bar{y}_s + (1 - f_s)\{(1 - c_s)\bar{y}_s + c_s\theta_0\} \tag{4.1}$$

where $c_s = \sigma_a^2/\{\sigma_a^2 + \nu(s)\sigma_0^2\}$. Two limiting cases of (4.1) are of interest: If $c_s = 0$, then $t_1^* = \bar{y}_s$ (compare Case A of Section 5.3), and if $c_s = 1$, then $t_1^* = f_s\bar{y}_s + (1 - f_s)\theta_0$ (compare Case C of Section 5.3).

The posterior of $\bar{y}$, given $d$, is $N(t_1^*, \sigma_p^2)$, where $\sigma_p^2 = (1 - f_s)[\sigma_a^2 + \{N - \nu(s)\}\sigma_1^2]/N$. Thus exact probability statements about $\bar{y}$ are easily computable.

A special case: Diffuse prior knowledge. In this case, $1/\sigma_0^2 = 0$, and $t_1^* = \bar{y}_s$. Given $d$, the quantity

$$\frac{\bar{y} - \bar{y}_s}{\{(1 - f_s)\sigma_a^2 / \nu(s)\}^{1/2}}$$

has the $N(0, 1)$ distribution. This result agrees with what is often recommended in classical survey sampling theory. ■

**Example 2.** Let $g(y_k \mid \theta, h)$ be $N(\theta, 1/h)$, where $\theta$ and $h$ are unknown. Then

$$g(\mathbf{y} \mid \theta, h) \propto h^{N/2} \exp\left\{-\frac{h}{2} \Sigma_1^N (y_k - \theta)^2\right\}$$

Assume that the prior knowledge of $\theta$ and $h$ can be summarized in the form of a normal-gamma distribution (see Raiffa and Schlaifer, 1961) with density

$$f(\theta, h \mid \boldsymbol{\phi}) \propto h^{\delta(n_0)/2} \exp\left\{\frac{-hn_0(\theta - \theta_0)^2}{2}\right\} h^{m_0/2-1} \exp\left\{\frac{-hm_0 a_0}{2}\right\}, \quad (4.2)$$

for $-\infty < \theta < \infty$, $0 < h < \infty$, where $\boldsymbol{\phi} = (n_0, \theta_0, m_0, a_0)$ is a vector of known parameters, $n_0 \geqslant 0, m_0 \geqslant 0, a_0 > 0$, and $\delta(n_0) = 0$ if $n_0 = 0$, $\delta(n_0) = 1$ if $n_0 > 0$.

Certain degenerate priors are admitted as special cases of (4.2): For example, if $n_0 = m_0 = 0$, (4.2) reduces to the diffuse degenerate prior

$$f(\theta, h \mid \boldsymbol{\phi}) = h^{-1} \qquad (-\infty < \theta < \infty, 0 < h < \infty)$$

which is uniform in $\theta$ and in $\log h$.

After simplification, the posterior of $(\theta, h)$ is obtained as

$$g(\theta, h \mid d, \boldsymbol{\phi}) \propto h^{1/2} \exp\left\{\frac{-hn_1(\theta - \theta_1)^2}{2}\right\} h^{m_1/2-1} \exp\left\{\frac{-hm_1 a_1}{2}\right\} \qquad (4.3)$$

for $0 < h < \infty$, $-\infty < \theta < \infty$, where $n_1 = n_0 + \nu(s)$, $m_1 = m_0 + \delta(n_0) + \nu(s) - 1$,

$$m_1 a_1 = m_0 a_0 + \Sigma_s (y_k - \bar{y}_s)^2 + \left\{\frac{\nu(s) n_0}{n_1}\right\} (\bar{y}_s - \theta_0)^2$$

and

$$\theta_1 = \frac{n_0 \theta_0 + \nu(s) \bar{y}_s}{n_0 + \nu(s)}$$

Using (3.3), observing that $\mu(\theta, h) = \theta$, we obtain the posterior expectation with respect to (4.3), $E_{\theta,h}(\theta \mid d, \boldsymbol{\phi}) = \theta_1$. Hence, from (3.4), we obtain the Bayes estimate

$$t_2^* = f_s \bar{y}_s + (1 - f_s)\{(1 - c_s)\bar{y}_s + c_s \theta_0\}$$

where $c_s = n_0/\{n_0 + \nu(s)\}$. (This solution is identical to $t_1^*$ of Example 1 if $n_0 = \sigma_a^2/\sigma_0^2$.)

The extreme cases of $c_s = 0$ and $c_s = 1$ are of some interest:

We have $c_s = 0$ if $n_0 = 0$, that is, if the prior variance of $\theta$, $1/hn_0$, is infinite, indicating that prior information about $\theta$ is absent. In this case $t_2^*$ reduces to $\bar{y}_s$.

We have $c_s = 1$ if $n_0 = \infty$. The prior variance of $\theta$ is then zero, indicating prior certainty that $\theta = \theta_0$. In this case $t_2^* = f_s\bar{y}_s + (1 - f_s)\theta_0$.

Hence, not surprisingly, $t_2^*$ agrees formally in these two extreme cases with the solutions of Cases A and C, respectively, of Section 5.3.

The posterior distribution of $\bar{y}$, given $d, \boldsymbol{\phi}$, is such that the quantity

$$\frac{\bar{y} - t_2^*}{[(1 - f_s)\{1/N + (1 - f_s)/n_1\}a_1]^{1/2}}$$

has Student's $t$-distribution with $m_1$ degrees of freedom. (The above results apply without modification if $n_0 = 0$ and/or $m_0 = 0$.)

A special case: Diffuse prior knowledge. Setting $n_0 = m_0 = 0$, we get $t_2^* = \bar{y}_s$. Given $d$, the quantity

$$\frac{\bar{y} - \bar{y}_s}{\{(1 - f_s)a_1/\nu(s)\}^{1/2}}$$

has Student's $t$-distribution with $\nu(s) - 1$ degrees of freedom, where

$$a_1 = \frac{\Sigma_s(y_k - \bar{y}_s)^2}{\nu(s) - 1}$$

The probability statements on $\bar{y}$ derived from this result again agree with classical procedures in survey sampling. The only difference *vis-à-vis* Example 1 is that $\sigma_a^2$ has been replaced by the estimate $a_1$. ∎

**Example 3.** Assume that auxiliary variable values $x_k > 0$ $(k = 1, \ldots, N)$ are available, and that the joint prior of $\mathbf{y}$, conditional on the unknown vector $\boldsymbol{\theta} = (\beta, h)$, is

$$f(\mathbf{y} \mid \boldsymbol{\theta}) \propto \left(\frac{h}{z_k}\right)^{N/2} \exp\left\{-h\,\Sigma_1^N \frac{(y_k - \beta x_k)^2}{2z_k}\right\}$$

and $z_k = v(x_k)$, for a known function $v(\cdot)$. Hence the $y_k$ are independent but not exchangeable. With minor modifications we can still use the analysis leading to the form (3.4) of the Bayes estimate. Assume the normal-gamma prior density for $\boldsymbol{\theta} = (\beta, h)$,

$$f(\boldsymbol{\theta} \mid \boldsymbol{\phi}) \propto h^{\delta(n_0)/2} \exp\left\{\frac{-hn_0(\beta - \beta_0)^2}{2}\right\} h^{m_0/2-1} \exp\left\{\frac{-hm_0a_0}{2}\right\} \quad (4.4)$$

for $-\infty<\beta<\infty$, $0<h<\infty$, where $\boldsymbol{\phi}=(n_0, \beta_0, m_0, a_0)$ is a vector of known parameters such that $n_0 \geqslant 0, m_0 \geqslant 0, a_0 > 0$, and $\delta(n_0)=0$ if $n_0=0, \delta(n_0)=1$ if $n_0>0$. If $n_0=m_0=0$, the diffuse degenerate prior $f(\boldsymbol{\theta} \mid \boldsymbol{\phi})=h^{-1}$ obtains.

After some computation, the posterior of $\boldsymbol{\theta}$, $f(\boldsymbol{\theta} \mid d, \boldsymbol{\phi})$, is found to be given by (4.4) provided $n_0, \beta_0, m_0, a_0$ are replaced, respectively, by $n_1, \beta_1, m_1$, and $a_1$, such that

$$n_1 = n_0 + \Sigma_s \frac{x_k^2}{z_k}$$

$$\beta_1 = \frac{n_0\beta_0 + (n_1 - n_0)\hat{\beta}}{n_1}$$

$$m_1 = m_0 + v(s) + \delta(n_0) - \delta(n_1)$$

$$m_1 a_1 = m_0 a_0 + \Sigma_s \frac{(y_k - \hat{\beta}x_k)^2}{z_k} + n_0 \left(1 - \frac{n_0}{n_1}\right)(\hat{\beta} - \beta_0)^2$$

and

$$\hat{\beta} = \frac{\Sigma_s x_k y_k / z_k}{\Sigma_s x_k^2 / z_k} \tag{4.5}$$

The Bayes estimate is constructed from (3.4), noting that

$${}_1\mathscr{E}_{\bar{s}\mid s}(\Sigma_{\bar{s}} y_k) = (\Sigma_{\bar{s}} x_k) E_{\theta}(\beta \mid d, \boldsymbol{\phi}) = (\Sigma_{\bar{s}} x_k)\beta_1$$

where $E_{\boldsymbol{\theta}}(\cdot \mid d, \boldsymbol{\phi})$ is with respect to $f(\boldsymbol{\theta} \mid d, \boldsymbol{\phi})$, the posterior of $\boldsymbol{\theta}$.

The Bayes estimate of $\bar{y}$ is, therefore,

$$t_3^* = f_s \bar{y}_s + (1 - f_s)\bar{x}_{\bar{s}}\{(1 - c_s')\hat{\beta} + c_s'\beta_0\} \tag{4.6}$$

where $\hat{\beta}$ is given by (4.5) and

$$c_s' = \frac{n_0}{n_0 + \Sigma_s x_k^2 / z_k}$$

Consider the two extreme cases: If $c_s'=0$, which occurs if $n_0=0$, that is, if the prior variance of $\beta$ is infinite, then (4.6) reduces to $t_3^* = f_s\bar{y}_s + (1-f_s)\bar{x}_{\bar{s}}\hat{\beta}$, which is identical to the Brewer–Royall estimate obtained from (4.1) in Section 5.4.

If $c_s'=1$, which happens if $n_0=\infty$, implying $\beta=\beta_0$ with prior probability one, then (4.6) becomes $t_3^* = f_s\bar{y}_s + (1-f_s)\bar{x}_{\bar{s}}\beta_0$.

The posterior distribution of $\bar{y}$, given $d$ and $\boldsymbol{\phi}$, is such that

$$\frac{\bar{y} - t_3^*}{[N^{-2}\{\Sigma_{\bar{s}} z_k + (\Sigma_{\bar{s}} x_k)^2 / n_1\} a_1]^{1/2}} \tag{4.7}$$

has Student's $t$-distribution with $m_1$ degrees of freedom.

A special case: Diffuse prior knowledge. Setting $n_0 = m_0 = 0$, we obtain the Brewer–Royall estimate, $t_3^* = f_s\bar{y}_s + (1 - f_s)\bar{x}_{\bar{s}}\hat{\beta}$. Probability statements on $\bar{y}$ can be derived from the fact that, given $d$ and $\boldsymbol{\phi}$, the quantity (4.7) with $n_1 = \Sigma_s x_k^2/z_k$ and

$$a_1 = \frac{\Sigma_s(y_k - \hat{\beta}x_k)^2/z_k}{\nu(s) - 1}$$

has Student's $t$-distribution with $\nu(s) - 1$ degrees of freedom.

If, furthermore, $z_k = v(x_k) = x_k$, then $t_3^* = \bar{x}\bar{y}_s/\bar{x}_s = t_R$, the traditional ratio estimator. The quantity

$$\frac{\bar{y} - t_R}{\{(1 - f_s)(\bar{x}\bar{x}_{\bar{s}}/\bar{x}_s)a_1/\nu(s)\}^{1/2}}$$

has Student's $t$-distribution with $\nu(s) - 1$ degrees of freedom, where

$$a_1 = \frac{\Sigma_s y_k^2/x_k - (\Sigma_s y_k)^2/(\Sigma_s x_k)}{\nu(s) - 1}$$

Compare Remark 1 of Section 5.4, and Section 5.5.

If $z_k = v(x_k) = x_k^2$, and $n_0 = m_0 = 0$, then $t_3^* = f_s\bar{y}_s + (1 - f_s)\bar{x}_{\bar{s}}\bar{r}_s$, where $\bar{r}_s = \Sigma_s r_k/\nu(s)$ and $r_k = y_k/x_k$. In this case,

$$\frac{\bar{y} - t_3^*}{[N^{-2}\{\nu(s)\Sigma_{\bar{s}}x_k^2 + (\Sigma_{\bar{s}}x_k)^2\}a_1/\nu(s)]^{1/2}}$$

has Student's $t$-distribution with $\nu(s) - 1$ degrees of freedom, where

$$a_1 = \frac{\Sigma_s(r_k - \bar{r}_s)^2}{\nu(s) - 1}$$

Compare Remark 2, Section 5.4. ∎

**Remark 1.** Examples 1–3 produce very appealing results by which probability statements on $\bar{y}$ are possible through normal and $t$-distributions. This is, of course, obtained only at the expense of having to give a highly detailed specification of the situation, including the normal-gamma prior, which simplifies the mathematics. Possibly, specification carried to such lengths may be seen by some as a weakness of the approach. ∎

**Remark 2.** The fiducial approach of Kalbfleisch and Sprott (1969) gives results which are identical to the "diffuse prior knowledge" cases of Examples 1–3 above. Assume again Model $E_{PI}$, where $g(y_k \mid \mu, \sigma^2)$ is known to be the $N(\mu, \sigma^2)$ density. For simplicity, let the effective sample size $\nu(s)$ be fixed at $n$.

If $\sigma^2$ is known, the resulting marginal fiducial distribution of $\bar{y}$, given the data, is such that

$$\frac{\bar{y} - \bar{y}_s}{\{(1 - f)\sigma^2/n\}^{1/2}}$$

is $N(0, 1)$; compare the identical result of Example 1 above.

If both $\mu$ and $\sigma^2$ are unknown, the resulting marginal fiducial distribution of $\bar{y}$ is derived as

$$f(\bar{y} \mid \bar{y}_s, v_y) = \int_\mu \int_{\sigma^2} g(\bar{y} \mid \mu, \sigma^2) f(\mu, \sigma^2 \mid \bar{y}_s, v_y)\, d\mu d\sigma^2$$

where the fiducial distribution $f(\mu, \sigma^2 \mid \bar{y}_s, v_y)$ has been given by Fisher (1956, p. 119); the pair $\bar{y}_s$ and $v_y = \Sigma_s(y_k - \bar{y})^2/(n - 1)$ is sufficient for $\mu, \sigma^2$. The resulting fiducial distribution is such that

$$\frac{\bar{y} - \bar{y}_s}{\{(1 - f)v_y/n\}^{1/2}}$$

has Student's $t$-distribution with $n - 1$ degrees of freedom; compare Example 2 above.

In an example involving an auxiliary variable, Kalbfleisch and Sprott's (1969) fiducial result coincides with the "diffuse prior knowledge" case of Example 3 above.

The fiducial approach of Kalbfleisch and Sprott (1969) should not be confused with the totally different fiducial framework of Godambe (1969b), Godambe and Thompson (1971b), which is "nonparametric," that is, no assumption of known superpopulation shape is made. ■

**Example 4.** Scott and Smith (1969b) considered the following two-stage sampling model:

The finite population consists of $N$ clusters. In the $i$th cluster there are $M_i$ units with associated values $y_{i1}, \ldots, y_{iM_i}$. At stage one, a sample $s$ of $n$ clusters is selected from the $N$ clusters. At stage two, a sample $s_i$ of $m_i$ distinct units is selected from the $i$th cluster selected at stage one.

Consider the following prior for the $y_{ik}$:

In the $i$th cluster, $y_{i1}, \ldots, y_{iM_i}$ are independent $N(\mu_i, \sigma_i^2)$ $(i = 1, \ldots, N)$. The means $\mu_1, \ldots, \mu_N$ are independent $N(\mu, \delta^2)$.

The resulting $\Sigma_1^N M_i$-variate normal prior for the $y_{ik}$-values is such that $\mathscr{E}(y_{ik}) = \mu$ for all $i, k$, and

$$\mathscr{C}(y_{ik}, y_{jl}) = \begin{cases} \delta^2 + \sigma_i^2, & i = j, k = l \\ \delta^2, & i = j, k \neq l \\ 0, & i \neq j \end{cases}$$

Suppose the linear parametric function

$$\omega = \Sigma_1^N \Sigma_1^{M_i} a_{ik} y_{ik}$$

is to be estimated, where the $a_{ik}$ are specified constants. As an estimate of $\omega$, we should, following the Bayesian reasoning in this section, use

$$\hat{\omega} = \Sigma_1^N \Sigma_1^{M_i} a_{ik} B_{ik} \tag{4.8}$$

where $B_{ik} = y_{ik}$ for $i \in s$, $k \in s_i$ (that is, when the unit $ik$ is included in the sample, so that its $y$-value is known exactly), and, for all other units, $B_{ik}$ equals the posterior mean of $y_{ik}$, given the data.

In the case of diffuse prior knowledge, the following results are obtained (for details, see Scott and Smith, 1969b):

(a) The prior of $\mu$ is uniform, and $\delta^2, \sigma_1^2, \ldots, \sigma_N^2$ are known. The appropriate values $B_{ik}$ in (4.8) are:

For units included in the sample:

$$B_{ik} = y_{ik}, \quad i \in s, \quad k \in s_i$$

For units not in the sample, but in a cluster from which some units have been sampled:

$$B_{ik} = \lambda_i \hat{\bar{y}}_i + (1 - \lambda_i)\hat{\bar{y}}, \quad i \in s, \quad k \notin s_i$$

where $\lambda_i = \delta^2/(\delta^2 + \sigma_i^2/m_i)$, $i \in s$, and $\hat{\bar{y}}_i = \Sigma_{s_i} y_{ik}/m_i$, $\hat{\bar{y}} = \Sigma_s \lambda_i \hat{\bar{y}}_i / \Sigma_s \lambda_i$;

For units from clusters not included in the sample $s$:

$$B_{ik} = \hat{\bar{y}}, \quad i \notin s$$

The posterior distribution of units not included in the sample is multivariate normal.

(b) Assuming $\sigma_i^2 = \sigma^2$ and $m_i = m$ $(i = 1, \ldots, N)$, let the diffuse prior for the unknown $\mu$, $\sigma^2$, and $\delta^2$ be

$$f(\mu, \sigma^2, \delta^2) \propto \lambda \sigma^{-2} \delta^{-2}$$

where $\lambda = \delta^2/(\delta^2 + \sigma^2/m)$, as discussed by Tiao and Tan (1965).

For each of the three categories of units, we should now use the following values $B_{ik}$ in (4.8)

$$B_{ik} = \begin{cases} y_{ik}, & i \in s, k \in s_i \\ \hat{\lambda}\hat{\bar{y}}_i + (1 - \hat{\lambda})\hat{\bar{y}}, & i \in s, k \notin s_i \\ \hat{\bar{y}}, & i \notin s \end{cases}$$

where

$$1 - \hat{\lambda} = \left\{ \frac{n(m-1)}{r(nm - m - 2)} \right\} \frac{I_x(c_1 + 1, c_2)}{I_x(c_1, c_2 + 1)}$$

with $c_1 = (n-1)/2$, $c_2 = (nm - m - 2)/2$, $r = n(m-1)x/(n-1)(1-x)$, $x = m\Sigma_s(\hat{\bar{y}}_i - \hat{\bar{y}})^2/\Sigma_s\Sigma_{s_i}(y_{ik} - \hat{\bar{y}})^2$, and $I_x(p, q)$ is the incomplete Beta function. As $\hat{\lambda}$ is difficult to compute in practice, the approximation $\hat{\lambda} = 1 - 1/r$ is suggested. ■

CHAPTER 7

# Efficiency Robust Estimation of the Finite Population Mean

Previous chapters have been theoretical in the sense of dealing mainly with the optimal choice of procedure given certain conditions and criteria. Often the established bestness property is a highly "local" phenomenon, that is, the optimality applies under conditions that are sometimes too severely restricted to meet practical exigencies. In the real world situation formal optimality may become secondary in importance to the pragmatic considerations. Even an inadmissible estimator may be highly useful in practice, provided it is a "good inadmissible" procedure.

For example, in discussing estimators in PPS sampling, Sampford (1975) lists three pragmatic criteria for choice between such methods: (i) ease of execution of the sampling procedure, (ii) ease of calculation of the joint inclusion probabilities $\alpha_{kl}$ needed in estimation of the variance of the estimate, (iii) stability of variance estimators. In view of these requirements the Horvitz–Thompson strategy $(ppsx, \bar{x}R_{yx})$, for example, may have to be rejected as impractical, even though it is optimal by certain mathematical criteria, and some sub-optimal alternative may be preferred.

## 7.1. THE PURPOSE OF ROBUSTNESS STUDIES IN SURVEY SAMPLING

The emphasis in this final chapter is on calculation of efficiencies of alternative strategies in situations where practical advantages of a certain strategy may outweigh its loss of efficiency relative to the best strategy. Comparison of efficiencies is also involved in robust estimation, another topic to be discussed in this chapter.

The study of robustness for finite populations is often simplified by recourse to superpopulation models. If a given strategy has been shown to be best for a specified model, then there is strong incentive to apply the strategy in practice, if real world conditions are deemed close to those of the model. But such a strategy can still be poor from a robustness point of view if it is highly sensitive to errors in the model. A prudent approach in many situations is, therefore, to choose a strategy that performs consistently well under a wide range of conditions (models), rather than one that is optimal under narrow conditions.

The general aim in robust estimation is to find an estimator or a strategy that "performs well" in some broad sense allowing for our uncertainty about the real world. Royall and Herson (1973a,b) discussed robustness in finite population estimation and suggested the following steps as a general guide in this type of research:

(i) If the assumed model were wrong, what circumstances were likely to exist;
(ii) describe these circumstances by an alternative model;
(iii) analyze the proposed procedure under the alternative model;
(iv) compare the proposed procedure with the preferred one under the alternative model.

Our discussion of robustness will be in the general spirit of points (i)–(iv).

The two main situations to be discussed are:

1. comparison of $p$-unbiased, or approximately $p$-unbiased, strategies,
2. comparison of strategies derived from the model-based approach, that is, when design aspects (such as $p$-unbiasedness) are considered unimportant.

In Sections 7.2–7.6 we shall study various available (approximately) $p$-unbiased strategies from a robustness point of view, under Model $G_R$ with $v(x) = x^g$. This model will be denoted Model $G_{Rg}$. The constant $g$ is usually unknown in practice, but one may have a good idea as to a plausible range for $g$. It is usually assumed that $0 \leq g \leq 2$. Experience indicates that this interval, or perhaps even the narrower interval $1 \leq g \leq 2$, would cover the majority of situations occurring in practice; see Cochran (1953, p. 212), Brewer (1963b). Since the true value of $g$ is difficult to ascertain, it is of interest to pinpoint strategies that perform well for all values of $g$ within a stated range.

Section 7.7 is devoted to the study of questions caused by the difficulty of maintaining $\xi$-unbiasedness under a variety of polynomial regression models. The problem can be controlled by the choice of design; this leads to the concept of balanced sampling, Royall and Herson (1973a).

## 7.2. USES OF AUXILIARY INFORMATION IN DESIGN-ORIENTED ESTIMATION

A question frequently arising in survey sampling concerns the utilization of known, positive auxiliary variable measurements $x_1, \ldots, x_N$, in an efficient as well as practical manner. In situations where a relationship is known to exist between the $y_k$ and the $x_k$, there is a strong possibility that the (labeled) numbers $x_1, \ldots, x_N$ can be used to construct more efficient strategies than $(srs, \bar{y}_S)$. It is known from Section 5.7 that $(srs, \bar{y}_S)$ is an optimal strategy, under certain formulations of the conditions, if the prior knowledge is symmetric with respect to the labels. In such cases, incorporation of auxiliary information attached to the labels does not improve on $(srs, \bar{y}_S)$.

When they do carry information, the auxiliary values $x_1, \ldots, x_N$ may be put to effective use either at the design stage, or at the estimation stage, or both. At the design stage, use of the auxiliary values would usually involve the choice of a sampling design such that the inclusion probabilities depend suitably on the $x_k$, as in the various PPS sampling designs. Several well known procedures use auxiliary information at the estimation stage. The classical ratio and regression estimators, usually used in conjunction with the design *srs*, are examples of this; that is, the estimators depend explicitly on $x_1, \ldots, x_N$.

The much publicized Horvitz–Thompson estimator is not necessarily the most efficient solution. The optimality results that we have established for the Horvitz–Thompson estimator are in effect too limited to be of much guidance. We showed in Section 3.5 that $t_{HT}$ can be claimed to be UMV $p$-unbiased estimator, but that was under restriction to a fixed FES($n$) design, that is, a fixed set of $\alpha_k$. We showed in Sections 4.4 and 4.5 that the strategy $(p_o, T_{HTo})$ was the minimum $\mathscr{E}$MSE strategy, but the restrictions in that case included a rather particular model formulation.

Moreover, in a multipurpose study one may have to estimate several fairly unrelated $y$-variables, while having access to several auxiliary $x$-variables. In this case it is necessary to strike some favorable medium with respect to utilization of the available auxiliary information, so that all the $y$-means can be estimated with reasonably good precision. If one single set of inclusion probabilities $\alpha_1, \ldots, \alpha_N$ is to be used, the Horvitz–Thompson estimator $t_{HT} = \Sigma_S y_k / N\alpha_k$ is not ideal, because it is hard to avoid that the $y_k$ and the $\alpha_k$ become weakly or negatively correlated for some of the $y$-variables. This would make $t_{HT}$ a poor estimator, as discussed in Remark 4 of Section 4.5.

We shall first define a system of (approximately or exactly) $p$-unbiased strategies for estimating $\bar{y}$ in which the $x_k$ enter at either or both of the two stages. The system incorporates most of the estimators discussed in earlier chapters and provides a fitting framework for the discussion to follow. Set

$$t_{HTy} = \Sigma_S y_k / N\alpha_k$$

$$t_{HTx} = \Sigma_S x_k / N\alpha_k$$

$$R_{yx} = \Sigma_S y_k / n x_k$$

where $\alpha_k > 0$ $(k = 1, \ldots, N)$ are the inclusion probabilities of any given FES($n$) design $p$. In general, $\alpha_k$ may be a function of $x_1, \ldots, x_N$. Moreover, let $h = h(\{(k, y_k, x_k, \alpha_k): k \in s\})$ be a specified function.

Consider estimators of $\bar{y}$ composed as

$$t = t_{HTy} + h(\bar{x} - t_{HTx}) \tag{2.1}$$

Many well-known, and some less-known but potentially interesting estimators belong to the system of estimators generated from (2.1) by various choices of $h$ and the $\alpha_k$. Table 7.1 gives some examples.

In the case where $\alpha_k = f = n/N$ $(k = 1, \ldots, N)$, there is reason to say that the auxiliary information has been ignored at the design stage. For example, this occurs for the design *srs*.

At the estimation stage, the auxiliary variable is, properly speaking, only utilized when $h \neq 0$. If $h = 0$, and the $\alpha_k$ are functions of the $x_k$, then the dependence of $t_{HTy}$ on the $x_k$ is induced by the choice of design, and not something that is introduced at the estimation stage.

Thus the case where $h = 0$ and $\alpha_k = f$ $(k = 1, \ldots, N)$ implies ignorance of the auxiliary information at *both* the design stage and the estimation stage. The estimator (2.1) is then the sample mean $\bar{y}_S$.

The case $h = \hat{\beta}$ indicates an estimated regression slope in a model of the relation between $y$ and $x$. For example, Model $G_R$ would justify the estimate

$$\hat{\beta} = \frac{\Sigma_S x_k y_k / v(x_k)}{\Sigma_S x_k^2 / v(x_k)} \tag{2.2}$$

More generally, one might use

$$\hat{\beta} = \frac{\Sigma_S (x_k - \bar{x}_S)(y_k - \bar{y}_S)/v(x_k)}{\Sigma_S (x_k - \bar{x}_S)^2 / v(x_k)}$$

In Table 7.1, the column $\alpha_k = f$ produces several of the classical estimators: The $p$-unbiased sample mean $\bar{y}_S$, the $p$-unbiased difference estimator $t_D = \bar{y}_S + c(\bar{x} - \bar{x}_S)$, where $c$ is a known constant, the approximately $p$-unbiased ratio estimator $t_R = \bar{x}\bar{y}_S/\bar{x}_S$ and the regression estimator $t_{REG} = \bar{y}_S + \hat{\beta}(\bar{x} - \bar{x}_S)$. The $p$-bias of the latter is known to be small under certain assumptions of a regression relationship between $y$ and $x$; see Cochran (1963, pp. 193–9). All of these estimators are consistent by Cochran's (1963, p. 20) definition of this term: When $n = N$, that is, when the sample consists of the whole population, the estimator becomes exactly identical to the population mean.

In generalization of these classical procedures we have the newer unequal probability sampling methods of the last two columns of Table 7.1. A basic question which we can at least answer partially in this chapter concerns the gains, if any, that the introduction of unequal probability sampling methods has meant in survey sampling.

For the design *ppsx*, that is, if $\alpha_k \propto x_k$, formula (2.1) gives the mean-of-the-ratios estimator, $\bar{x}R_{yx}$, regardless of the value of $h$.

**Table 7.1. Estimators of the Type $t_{HTy} + h(\bar{x} - t_{HTx})$.**

| Estimation Stage: Choice of $h$ | Design Stage: Choice of $\alpha_k$ | | |
|---|---|---|---|
| | $\alpha_k = f$ | $\alpha_k \propto x_k$ | Arbitrary $\alpha_k$ |
| $h = 0$ | $\bar{y}_S$ | $\bar{x}R_{yx}$ | $t_{HTy}$ |
| $h = c$ | $t_D = \bar{y}_S + c(\bar{x} - \bar{x}_S)$ | $\bar{x}R_{yx}$ | $t_{GD} = t_{HTy} + c(\bar{x} - t_{HTx})$ |
| $h = t_{HTy}/t_{HTx}$ | $t_R = \bar{x}\bar{y}_S/\bar{x}_S$ | $\bar{x}R_{yx}$ | $t_{GR} = \bar{x}t_{HTy}/t_{HTx}$ |
| $h = \hat{\beta}$ | $t_{REG} = \bar{y}_S + \hat{\beta}(\bar{x} - \bar{x}_S)$ | $\bar{x}R_{yx}$ | $t_{GREG} = t_{HTy} + \hat{\beta}(\bar{x} - t_{HTx})$ |

By substitution of arbitrary inclusion probabilities for the constant ones in the column $\alpha_k = f$ of Table 7.1, we obtain what may be termed a generalized sample mean, namely, the Horvitz–Thompson estimator $t_{HTy}$. The same analogy produces the other estimators of the final column of Table 7.1: The generalized difference estimator $t_{GD} = t_{HTy} + c(\bar{x} - t_{HTx})$, the generalized ratio estimator $t_{GR} = \bar{x}t_{HTy}/t_{HTx}$ suggested by Brewer (1963b), Sukhatme and Sukhatme (1970), and finally the generalized regression estimator $t_{GREG} = t_{HTy} + \hat{\beta}(\bar{x} - t_{HTx})$ suggested by Cassel, Särndal, and Wretman (1976) and discussed in Sections 5.4 and 7.6 for the special case where $\hat{\beta}$ is given by (2.2) and $\alpha_k \propto \sqrt{v(x_k)}$.

While $t_{HTy}$ and $t_{GD}$ are $p$-unbiased, $t_{GR}$ and $t_{GREG}$ are in general not. The $p$-bias of $t_{GR}$ tends to vanish as $n$ increases; that of $t_{GREG}$ can be expected to be small when the $y_k$ and the $x_k$ have a linear regression relationship. By Cochran's definition of consistency, $t_{HTy}$, $t_{GD}$, $t_{GR}$, and $t_{GREG}$ are consistent. Little is known about $t_{GR}$ and $t_{GREG}$; they remain interesting possibilities for the multipurpose survey where several $y$ means have to be estimated using one fixed set of unequal inclusion probabilities $\alpha_1, \ldots, \alpha_N$. (As far as $t_{GR}$ goes, the similarity with Hájek's estimator (4.10) of Section 5.4 should be noted.)

## 7.3. MEAN-OF-THE-RATIOS STRATEGIES AND RATIO STRATEGIES

Among the types of strategies listed in Table 7.1, we consider in particular the PPS-sampling strategy $(ppsx, \bar{x}R_{yx})$ and the classical ratio estimation strategy $(srs, t_R)$. Given certain conditions, each in its own way makes efficient use of auxiliary information, but unfortunately neither strategy has been found ideal: The former has an intractable design (see Section 1.4); the latter is $p$-biased. As a consequence, the literature suggests a number of variations on each of the two themes.

Strategies of the first type will be called $M$-strategies (where $M$ stands for Mean-of-the-ratios) and those of the second type will be called $R$-strategies (where $R$ stands for Ratio). The various $M$- and $R$-strategies represent a body of well-researched procedures. Material reported in Sections 7.4–7.6 has the character of a case study. It illustrates the methods currently in use for comparison of sampling strategies, and is limited by our choice of the regression model, Model $G_R$, as a point of departure for making the comparisons.

An $M$-strategy will be denoted $M = (p, t)$, where the design usually involves PPS sampling and where the estimator usually combines the ratios $y_k/x_k$, for $k \in S$, into a (possibly weighted) average. An $R$-strategy will be denoted $R = (p, t)$. The design $p$ is usually simple random sampling with or without replacement, and the estimator is either the classical ratio estimator $t_R = \bar{x}\bar{y}_S/\bar{x}_S$ or some modified version thereof.

The notation will be as follows for five $M$-strategies to be considered:

1. The Horvitz–Thompson (1952) strategy,

$$M_{HT} = (ppsx, \bar{x}R_{yx})$$

2. The Hansen–Hurwitz (1943) strategy (repeats allowed),

$$M_{HH} = (ppsrx, \bar{x}R^{\circ}_{yx})$$

3. The Raj (1956) strategy,

$$M_{DR} = (ppsux, t_{DR})$$

4. The Murthy (1957) strategy,

$$M_{MU} = (ppsux, t_{MU})$$

The designs in the above strategies were defined in Section 1.4. The estimators $\bar{x}R^{\circ}_{yx}$, $\bar{x}R_{yx}$, and $t_{DR}$ were defined in Section 1.6; finally, $t_{MU}$ was given by (3.4) of Section 2.3. Repeated labels are possible only for the strategy $M_{HH}$; in this case, $R^{\circ}_{yx}$ is the average of the $n$ ratios $y_k/x_k$ for $k \in \mathbf{S}$, repeats included. In the remaining three strategies, the estimators are based on exactly $n$ distinct labels, since the designs are FES($n$). The comparison is somewhat biased against $M_{HH}$, in the sense that this strategy has the expected effective sample size $E\{\nu(\mathbf{S})\} \leqslant n$. The final $M$-strategy is:

5. The Rao–Hartley–Cochran (1962) strategy,

$$M_{RHC} = (ppsgx, t_{RHC})$$

where $ppsgx$ denotes the FES($n$) design consisting in drawing, by PPS of $x$, *one* label from each of $n$ pairwise disjoint and collectively exhaustive subsets $\mathscr{U}_i$ of lables into which $\{1, \ldots, N\}$ has been divided at random. That is, from the subset $\mathscr{U}_i$ containing $N_i$ units, draw the label $k_i$ with probability $x_{k_i}/N_i\bar{x}_i$, where $\bar{x}_i = \Sigma_{\mathscr{U}_i} x_k/N_i$. Assuming $k_i$ to be the label drawn from $\mathscr{U}_i$, the estimator is

$$t_{RHC} = \sum_{i=1}^{n} \frac{y_{k_i} N_i \bar{x}_i}{N x_{k_i}}$$

(Rao, Hartley and Cochran (1962) showed that it is optimal to choose the number of units in each subset $\mathscr{U}_i$ as nearly equal as possible, that is, equal to either $m$ or $m+1$, where $m$ is the integer part of $N/n$. With no serious loss of generality, our discussion of the $M_{RHC}$ strategy will assume that $N/n$ is integer.)

**Remark 1.** All of the above $M$-strategies are $p$-unbiased. On theoretical grounds the strategy $M_{HT}$ would sometimes be preferred to the others. Under Model $G_R$, for example, Theorems 4.1 and 4.2 state that $M_{HT}$ is the best $p$-unbiased estimator if $v(x) = x^2$. But if, under the same model, $v(x) = x^g$ with $g \neq 2$, no optimality results are available. As a consequence, much research has concentrated on comparing various ad hoc strategies available in the literature.

In Section 1.4 we commented on the serious practical disadvantages associated with the strategy $M_{HT}$, namely, the absence of an easy way to execute an FES($n$) design with given inclusion probabilities $\alpha_k \propto x_k$. Hence in practice one may prefer one of the other $M$-procedures, even though the estimator may be inadmissible, as in the case of $M_{HH}$; see Example 2, Section 2.3. In fact, $M_{HH}$ is frequently used in practice because the design *ppsrx* is simpler to implement.

Thus it becomes of interest to compute the loss of efficiency of $M_{HH}$ relative to $M_{HT}$. While negligible if the population size $N$ is very large, the loss can be considerable for small and moderate size $N$, as we shall see in Sections 7.4–7.5.

In contrast to $M_{HH}$, the strategies $M_{DR}$, $M_{MU}$, and $M_{RHC}$ avoid repeats. The estimator $t_{MU}$ is complicated if $n > 2$ but small sample results indicate that $M_{MU}$ is frequently better than the other $M$-strategies under consideration. Yet little is known about the properties of $M_{MU}$ beyond the fact that its variance (and hence its expected variance under any model) is always less than that of $M_{DR}$, as shown in Section 2.3.

The design *ppsux* has the disadvantage of requiring recalculation of the drawing probabilities from one draw to the next. By contrast, the *ppsgx* design used in $M_{RHC}$ offers an elegant way of avoiding both repeats and recalculation of probabilities. ■

The notation will be as follows for the five $R$-strategies to be discussed. (The sample size is assumed to be fixed at $n$ for the design *srs*. For the design *srsr*, $n$ is the number of with replacement draws):

1. The classical ratio estimation strategy,

$$R_C = (srs, t_R)$$

where $t_R = \bar{x}\bar{y}_S/\bar{x}_S$,

2. The with replacement version of $R_C$,

$$R_{CR} = \left( srsr, \frac{\bar{x}\bar{y}_{\mathrm{S}}}{\bar{x}_{\mathrm{S}}} \right)$$

where $\bar{y}_{\mathrm{S}} = \Sigma_{\mathrm{S}} y_k/n$, $\bar{x}_{\mathrm{S}} = \Sigma_{\mathrm{S}} x_k/n$ (repeats included),

3. The Tin (1965) strategy,

$$R_{TI} = (srs, t_{TI})$$

where

$$t_{TI} = t_R \left( 1 + \frac{c_{yx}/\bar{x}_S\bar{y}_S - v_x/\bar{x}_S^2}{n} \right)$$

and

$$c_{yx} = \frac{\Sigma_S(x_k - \bar{x}_S)(y_k - \bar{y}_S)}{n-1}; \quad v_x = \frac{\Sigma_S(x_k - \bar{x}_S)^2}{n-1}$$

4. The strategy of Hájek (1949), Lahiri (1951), Midzuno (1952), and Sen (1953),

$$R_{LA} = (pps\Sigma x, t_R)$$

where $pps\Sigma x$ denotes the FES($n$) design such that

$$p(s) = \frac{\Sigma_s x_k}{\binom{N-1}{n-1} \Sigma_1^N x_k} \tag{3.1}$$

for any $s$ with $\nu(s) = n$. (In the literature, the procedure is often associated with the name of Lahiri, hence our abbreviation $R_{LA}$.) Hájek (1949), Midzuno (1952), and Sen (1953) gave a simple mechanism for implementing the design $pps\Sigma x$: Draw the first label $k_1$ with PPS of $x$, that is, with probability $x_{k_1}/\Sigma_1^N x_k$. Then, without replacing the first unit, use the design $srs$ for drawing $n-1$ labels from the remaining $N-1$.

5. The Hartley–Ross (1954) strategy,

$$R_{HR} = (srs, t_{HR})$$

where

$$t_{HR} = \bar{x}R_{yx} + \left\{ \frac{n(N-1)}{N(n-1)} \right\} (\bar{y}_S - \bar{x}_S R_{yx})$$

**Remark 2.** Among the $R$-strategies just mentioned, $R_{LA}$ and $R_{HR}$ are exactly $p$-unbiased, while $R_C$, $R_{CR}$ and $R_{TI}$ are only approximately so. As is known from Section 5.4, the predictor $T_R$ minimizes $\mathscr{E}(T - \bar{Y})^2$ among linear $\xi$-unbiased predictors, under Model $G_{R1}$, for any given $s$. However, it is also well known, for example, from Cochran (1963, p. 160) that $R_C$ is approximately (to order $n^{-1}$) $p$-unbiased. We have included the approximately $p$-unbiased version $R_{CR}$ mainly to see how much efficiency is lost because of with replacement sampling.

Several $R$-strategies have been proposed as a result of efforts to reduce or eliminate the $p$-bias of $R_C$, while preferably keeping the attractive design *srs*. Thus the estimator has to be an adjusted form of $t_R$. Among these are procedures suggested by Quenouille (1956), which uses the jackknifing technique, Mickey (1959), Pascual (1961), and Tin (1965). Of these we have only listed Tin's strategy $R_{TI}$. From an efficiency point of view, it appears that there are no great differences among these attempts; see Rao and Rao (1971a). We have selected $R_{TI}$ as it usually performs somewhat better than the others.

The other obvious possibility for achieving $p$-unbiasedness is to keep the estimator $t_R$ while adjusting the design. The strategy $R_{LA}$ does this, and the mechanism given above for implementing the design $pps\Sigma x$ is reasonably simple. Sen (1953) gave the inclusion probabilities for the design $pps\Sigma x$ as

$$\alpha_k = \frac{(N-n)x_k/N\bar{x} + (n-1)}{N-1}$$

which agrees closely, when $f = n/N$ is small, with the inclusion probabilities $\alpha_k = x_k/N\bar{x}$ of the design *ppsx*, even though $pps\Sigma x$ requires only one PPS draw out of $n$.

The Hartley–Ross strategy $R_{HR}$ is a mixture of features of $R$- and $M$-procedures: The design is *srs* but the estimator is rather of $M$-type. The first term, $\bar{x}R_{yx}$, of the estimator $t_{HR}$ is obviously $p$-biased (unless all $x_k$ are equal) under the design *srs*. This $p$-bias under *srs* is, however, removed by adding the second term of $t_{HR}$. Note that $t_{HR}$ is a special case of the Hájek (1959) estimator (4.11) of Section 5.4, namely the one obtained if $a_k = 1$ $(k = 1, \ldots, N)$. The Hartley–Ross technique was extended by Lanke (1975), who constructed a $p$-unbiased estimator based on removal of the $p$-bias of the strategy $(p, \bar{x}R_{yx})$ for an arbitrary FES($n$) design $p$. ∎

Several possibilities exist for comparison of strategies:

1. The theoretical approach: To establish, between the variances (or expected variances) of two strategies, an inequality which holds under some general condition. Sometimes the theoretical approach is based on approximations. Especially in the case of ratio strategies, approximative variances

obtained from Taylor expansions are usually used in place of the mathematically intractable exact expressions. Another technique involves replacement of the finite population by an approximating infinite population of specified shape. Strategies can then be compared in terms of the approximative but simplified variances.

2. Empirical or semitheoretical approaches: Computation of the variances (or expected variances) of competing strategies for actual finite populations, natural or artificial. For a description of these approaches, see Section 7.5.

If $H = (p, t)$ denotes any specified strategy, for example, an $M$-strategy or an $R$-strategy, we shall denote $\text{MSE}(p, t) = E(t - \bar{y})^2$ by $\text{MSE}(H)$. When MSE is used to compare two strategies for a given population, a strategy is called more efficient (less efficient) according as its MSE is less (greater) than that of its competitor.

An often simpler way to compare strategies is by means of their average MSE over a series of populations generated under a superpopulation model. For a given superpopulation model, we shall write $\mathscr{E}\text{MSE}(H)$ for $\mathscr{E}\text{MSE}(p, T) = \mathscr{E}E(T - \bar{Y})^2$, where $T$ is the predictor obtained from the estimator $t$ by replacing $y_k$ by $Y_k$ for $k \in S$. When expected MSE is involved, we shall use the quantity

$$A(H_1; H_2) = \frac{\mathscr{E}\text{MSE}(H_2)}{\mathscr{E}\text{MSE}(H_1)} - 1 \tag{3.2}$$

as the formal basis for comparing two strategies $H_1$ and $H_2$ for a given population. Note that $A(H_1; H_2)$ is not symmetric: $A(H_1; H_2) \neq A(H_2; H_1)$ unless both are zero. If $A(H_1; H_2) > 0$, $H_1$ is said to be more efficient than $H_2$ and $A(H_1; H_2)$ is called the *efficiency gain* of $H_1$ over $H_2$. If $A(H_1; H_2) < 0$, $H_1$ is said to be less efficient than $H_2$, and $|A(H_1; H_2)|$ is called the *efficiency loss* of $H_1$ relative to $H_2$. Finally, if $A(H_1; H_2) = 0$, $H_1$ and $H_2$ will be called *equally efficient.*

## 7.4. COMPARISONS AMONG *M*- AND *R*-STRATEGIES, THEORETICAL APPROACHES

Under the subheadings **A**, **B**, and **C** we present some results on expected MSE's for $M$- and $R$-strategies. Most comparisons in Sections 7.4–7.6 will be performed under Model $\text{G}_{Rg}$, which is our notation for Model $\text{G}_R$ with $v(x) = x^g$; see Table 4.1. As required by this model, the auxiliary variable measurements $x_1, \ldots, x_N$ are known but arbitrary positive numbers. If $H_1$ and $H_2$ are strategies such that $\mathscr{E}\text{MSE}(H_1) < \mathscr{E}\text{MSE}(H_2)$ under Model $\text{G}_{Rg}$, it means that $H_1$ is uniformly more efficient than $H_2$ in the sense that the inequality holds for all $\mathbf{x} = (x_1, \ldots, x_N)$ such that $x_k > 0$ for $k = 1, \ldots, N$. For simplicity we rule out the uninteresting case where all $x_k$ are equal, and assume also that $n > 1$. Set

$$p_k = \frac{x_k}{\Sigma_1^N x_k} \tag{4.1}$$

and, for any nonnegative constant $a$,

$$m_a = \frac{\Sigma_1^N x_k^a}{N}. \tag{4.2}$$

In addition to other factors, the magnitude of efficiency gains or losses depends heavily on the population size $N$ and on the sampling fraction $f = n/N$. Results for small sample sizes, even as extreme as $n = 2$, are also of interest, for example, for situations where stratification is carried to a point where one would only want to select a few units from a stratum.

### A. Some Exact Inequalities

The following theorem and its corollary deal with $M_{HT}$, $M_{HH}$, and $M_{RHC}$. The results are due in part to J. N. K. Rao (1966b) and in part to Foreman and Brewer (1971).

**Theorem 7.1.** *Under Model $G_{Rg}$, the M-strategies $M_{HT}$, $M_{HH}$, and $M_{RHC}$ satisfy the following relations:*

$$\mathscr{E}V(M_{HH}) = \frac{(m_{g-1}m_1 - m_g/N)\sigma^2}{n}$$

$$\mathscr{E}V(M_{HH}) - \mathscr{E}V(M_{RHC}) = \frac{(n-1)\,\mathscr{E}V(M_{HH})}{N-1} > 0$$

$$\mathscr{E}V(M_{HH}) - \mathscr{E}V(M_{HT}) = \frac{(n-1)m_g\sigma^2}{nN} > 0$$

*where $m_g$ is defined by (4.2). In the second formula, the ξ-expectation operator may be deleted, that is, the formula also applies to the p-variances.*

**Proof.** We have

$$V(M_{HH}) = \frac{(\Sigma_1^N Y_k^2/p_k)/N^2 - \bar{Y}^2}{n}$$

where $p_k$ is given by (4.1). Averaging with respect to the model we obtain $\mathscr{E}V(M_{HH})$, as given in the theorem. Secondly, it is easily seen that

$$V(M_{RHC}) = \left(\frac{N-n}{N-1}\right) V(M_{HH})$$

Finally,

$$V(M_{HT}) = \frac{\Sigma_1^N Y_k^2/\alpha_k + \Sigma\Sigma_{k \neq l}^N (Y_k Y_l/\alpha_k \alpha_l)\alpha_{kl}}{N^2} - \bar{Y}^2 \tag{4.3}$$

where $\alpha_k = np_k$ $(k = 1, \ldots, N)$. Averaging with respect to Model $G_{Rg}$ and using that $\Sigma\Sigma_{k \neq l}^N \alpha_{kl} = n(n-1)$, we get

$$\mathscr{E}V(M_{HT}) = \frac{(m_{g-1}m_1 - fm_g)\sigma^2}{n} \tag{4.4}$$

which gives the third statement of the theorem. □

**Corollary 7.1.** *Under Model $G_{Rg}$,*

$$\mathscr{E}V(M_{RHC}) - \mathscr{E}V(M_{HT}) = \frac{\{(n-1)/(N-1)\}(m_g - m_{g-1}m_1)\sigma^2}{n}$$

*which is $\gtreqless 0$ according as $g \gtreqless 1$.*

**Proof.** Theorem 7.1 gives the desired difference between the $\mathscr{E}V$'s of $M_{RHC}$ and $M_{HT}$. Moreover, $m_g - m_{g-1}m_1 > 0$, $= 0$, or $< 0$ according as the correlation between the positive numbers $x_k^{g-1}$ and $x_k$ across the labels is $> 0$, $= 0$, or $< 0$, that is, according as $g$ is $> 1$, $= 1$, or $< 1$. □

**Remark 1.** Theorem 7.1 and Corollary 7.1 show that $M_{HT}$, $M_{HH}$, and $M_{RHC}$ are equally efficient, from the $\mathscr{E}V$ point of view, when $N \to \infty$, provided $m_g$, $m_{g-1}$, and $m_1$ remain bounded. For moderate size populations little is known about the efficiency loss of $M_{HH}$ and $M_{RHC}$ relative to $M_{HT}$; it is possible that the losses are rather small. When $N$ is small, and $f = n/N$ relatively large, the efficiency losses can be considerable, see Section 7.5. ■

**Remark 2.** As shown by J. N. K. Rao (1966b), $M_{MU}$ is uniformly more efficient than $M_{HT}$ when $g = 1$ in Model $G_{Rg}$. For $1 < g < 2$, no uniformly valid conclusion is possible: The comparison depends on the configuration of the $x_k$. For $g = 2$, $M_{HT}$ is, by virtue of Theorem 4.2, uniformly more efficient than any other strategy. The problem with $M_{MU}$ in the case $g = 1$ is that, although it is more efficient than most existing alternatives, it has no known bestness properties. ■

**Remark 3.** An example from T. J. Rao (1972) shows that $M_{MU}$ can be improved upon: For $n = 2$ and for all values of $g$, the $p$-unbiased strategy $(p_1, m_{t1}R_{yt})$ has smaller $\mathscr{E}V$ than $M_{MU}$ under Model $G_{Rg}$, where (i) the design $p_1$ has the inclusion probabilities $\alpha_k \propto t_k = x_k^{g/2}$ *and* satisfies $\Sigma_S x_k/t_k = \text{constant}$; (ii) $m_{t1}R_{yt} = (\Sigma_1^N t_k)(\Sigma_S y_k/t_k)/Nn$ is the Horvitz–Thompson estimator for the design $p_1$. It would obviously be difficult to implement the design $p_1$. ■

As will be shown later, $R_{LA}$ is by and large the most efficient of the $R$-strategies listed in Section 7.3. The following theorem, due to J. N. K. Rao (1966b), makes an exact comparison between $R_{LA}$ and $M_{HT}$.

**Theorem 7.2.** *Under Model* $G_{Rg}$, $\mathscr{E}V(R_{LA}) \gtreqless \mathscr{E}V(M_{HT})$ *according as* $g \gtreqless 1$.

**Proof.** Let $s_i$, $i = 1, 2, \ldots, M = \binom{N}{n}$, be an enumeration of the sets $s$ of fixed size $n$. For any constant $a$, set

$$W_{ai} = \Sigma_{k \in s_i} Y_k^a; \quad u_{ai} = \Sigma_{k \in s_i} x_k^a$$

We obtain first

$$V(R_{LA}) = \frac{m_1 \left( \sum_{i=1}^{M} W_{1i}^2 / u_{1i} \right)}{nM} - \bar{Y}^2$$

Averaging with respect to Model $G_{Rg}$ and using the identity

$$nm_a = \sum_{i=1}^{M} \frac{u_{ai}}{M} \tag{4.5}$$

for $a = 1$, we get

$$\mathscr{E}V(R_{LA}) = \frac{\left\{ m_1 \left( \sum_{i=1}^{M} u_{gi}/u_{1i} \right) \Big/ M - fm_g \right\} \sigma^2}{n}$$

This is to be compared with $\mathscr{E}V(M_{HT})$ given by (4.4). Using (4.5) for $a = g - 1$, we have

$$\mathscr{E}V(R_{LA}) - \mathscr{E}V(M_{HT}) = \frac{m_1 \left( \sum_{i=1}^{M} d_i/u_{1i} \right)}{nM}$$

where $d_i = u_{gi} - u_{g-1,i} u_{1i}/n$. But for $i = 1, \ldots, M$ we have $d_i \gtreqless 0$ according as the within-$s_i$ correlation between $x_{g-1}$ and $x$ is $\gtreqless 0$, that is, according as $g \gtreqless 1$. □

## B. Some Approximative Inequalities

The mathematical treatment of ratio procedures usually leads to cumbersome mathematical expressions. Frequently, one has to resort to approximations valid for large $N$ and/or $n$, as in the results now to be discussed.

Recall first the approximations, valid for "large $n$," given, for example, in Cochran (1963),

$$\mathrm{MSE}(R_{CR}) \simeq \frac{V}{n} \tag{4.6}$$

$$\mathrm{MSE}(R_C) \simeq \frac{(1-f)V}{n} \tag{4.7}$$

where

$$V = \frac{\Sigma_1^N (Y_k - x_k \bar{Y}/\bar{x})^2}{N-1}$$

Taking the $\xi$-expectation of (4.6) and (4.7) under Model $G_{Rg}$ we obtain, as Foreman and Brewer (1971),

$$\mathscr{E}\mathrm{MSE}(R_{CR}) \simeq \frac{\{m_g + (m_g m_2/m_1^2 - 2m_{g+1}/m_1)/N\}\sigma^2}{n} \tag{4.8}$$

and

$$\mathscr{E}\mathrm{MSE}(R_C) \simeq \left(\frac{N-n}{N-1}\right) \mathscr{E}\mathrm{MSE}(R_{CR}) \tag{4.9}$$

where $m_g$ is given by (4.2).

**Remark 4.** The loss of efficiency due to sampling with replacement is about the same in $M$-procedures as in $R$-procedures. From Theorem 7.1, $A(M_{HH}; M_{RHC}) = -(n-1)/(N-1) \simeq -f$. From (4.8) and (4.9), $A(R_{CR}; R_C) \simeq -f$. (The efficiency loss $(n-1)/(N-1)$ also obtains in the comparison $(srsr, \bar{y}_S)$ relative to $(srs, \bar{y}_S)$.) In all three cases, the efficiency losses are the same whether we measure in terms of MSE's or expected MSE's. ■

**Remark 5.** The following approximative comparisons offer a perspective on $M$-strategies compared to $R$-strategies under Model $G_{Rg}$:

(i) Using Theorem 7.1 and formula (4.8), dropping the terms of $0(N^{-1})$,

$$\mathscr{E}\mathrm{MSE}(R_{CR}) - \mathscr{E}V(M_{HH}) \simeq \frac{(m_g - m_{g-1}m_1)\sigma^2}{n}$$

(ii) Using Theorem 7.1 and formula (4.9),

$$\mathscr{E}\mathrm{MSE}(R_C) - \mathscr{E}V(M_{RHC}) \simeq \frac{(1-f)(m_g - m_{g-1}m_1)\sigma^2}{n}$$

(iii) From Corollary 7.1,

$$\mathscr{E}V(M_{RHC}) - \mathscr{E}V(M_{HT}) \simeq \frac{f(m_g - m_{g-1}m_1)\sigma^2}{n}$$

Appearing in all three differences, the factor $m_g - m_{g-1}m_1$ is $\gtreqless 0$ according as $g \gtreqless 1$; see Corollary 7.1. The result (ii) was obtained under an almost equivalent model by Avadhani and Sukhatme (1970). Moreover, (ii) also applies in the comparison between $R_{LA}$ and $R_{RHC}$ (J. N. K. Rao, 1966b), since $\mathscr{E}V(R_{LA}) = \mathscr{E}\text{MSE}(R_C)$ when $n$ is large; see Section 7.6. Adding the results (ii) and (iii), and comparing with (i), we see that the absolute reduction (if $g > 1$) in $\mathscr{E}$MSE is approximately the same, $(m_g - m_{g-1}m_1)\sigma^2/n$, in going from $R_C$ to $M_{HT}$ as it is in going from $R_{CR}$ to $M_{HH}$, as shown by Foreman and Brewer (1971). If the sampling fraction $f = n/N$ is negligible, we get Cochran's (1953, p. 211) result that the $\mathscr{E}$MSE's of $R_C$ and $M_{HH}$ differ by approximately $(m_g - m_{g-1}m_1)\sigma^2/n$. We can expect that important efficiency differences may exist between $R$-strategies and $M$-strategies if $g \neq 1$, that is, if $m_g \neq m_{g-1}m_1$. A quantitative assessment of these differences will be given in Section 7.6. ■

### C. Comparing $R$-Strategies for Gamma($r$) Populations

We conclude this section by a comparison of $R$-strategies. Rao and Rao (1971a,b) considered simple random sampling from infinite populations under the model $y_i = \alpha + \beta x_i + e_i$ $(i = 1, \ldots, n)$, $\mathscr{E}(e_i \mid x_i) = 0$, $\mathscr{V}(e_i \mid x_i) = \sigma^2 x_i^g$, $\mathscr{C}(e_i, e_j \mid x_i, x_j) = 0$ $(i \neq j)$, and $x$ has the gamma distribution with parameter $r$. The strategies involved in the comparison were $R_C$, $R_{TI}$, $R_{HR}$ and the above-mentioned approximately $p$-unbiased $R$-strategies of Quenouille (1956) and Mickey (1959). Setting $\alpha = 0$ we obtain certain conclusions valid under Model $G_{Rg}$ and under the gamma($r$) population assumption. (The gamma($r$) approximation is discussed in detail in the following section.) These assumptions make it possible to express the $\mathscr{E}$MSE of each strategy as a function of $r, g$, and the sample size $n$. Rao and Rao (1971a,b) evaluated this function for $r = 1, 2, 3, 4$; $g = 0, 1, 2$; $n = 4, 6, 8, 12$. Some specific conclusions were:

1. For all $r$ and $n$, the classical $R$-strategy $R_C$ was more efficient than its competitors for $g = 2$, slightly more efficient for $g = 1$, while for $g = 0$, $R_{TI}$ was more efficient than $R_C$.
2. The Tin strategy $R_{TI}$ was second best for $g = 1$ and 2. For $g = 2$, the efficiency loss of $R_{TI}$ relative to $R_C$ ranged from 17% $(n = 4, r = 1)$ to 4% $(n = 12, r = 4)$. For $g = 1$, $R_{TI}$ lost a few percent at most relative to $R_C$. On the other hand, when $g = 0$, $R_{TI}$ showed efficiency gains over $R_C$ ranging from 38% $(n = 4, r = 1)$ to 4% $(n = 12, r = 4)$.
3. The Hartley–Ross strategy $R_{HR}$, not evaluated for all parameter combinations, was in general somewhat less efficient than $R_{TI}$ for $g = 0$, and about equally efficient as $R_{TI}$ for $g = 1, 2$.
4. The Quenouille (1956) and Mickey (1959) strategies were found to be close in efficiency to $R_{TI}$, except if $n$ and $r$ are extremely small. Thus $R_{TI}$ gained about 30% in efficiency over the Quenouille and Mickey strategies for $g = 2, n = 4, r = 1$.

In summary, the efficiency loss of the worst of the five compared $R$-strategies compared to the best one tended, for every $g$, to decrease toward zero as $n$ increases and as $r$ increases. For $n = 12$, $r = 4$, the least efficient strategy lost only a few percent relative to the best one.

To this we can add that $R_{LA}$ is fractionally more efficient than $R_C$, see Remark 2 of Section 7.6. In fact $R_{LA}$, $R_C$, and $R_{HR}$ have the same $\mathscr{E}$MSE to order $n^{-1}$, namely, $\{\Gamma(r+g)/\Gamma(r)\}\sigma^2/n$; see P. S. R. S. Rao (1968, 1969) and Section 7.5. We conclude that no great differences in efficiency exist among $R$-strategies, at least not in the case of gamma populations and as long as $n$ is not extremely small.

## 7.5. COMPARISONS AMONG $M$-STRATEGIES, SEMITHEORETICAL AND EMPIRICAL APPROACHES

A drawback with comparisons based on inequalities, as in most of the previous section, is that conclusions do not go beyond statements of the type "$M_0$ is more efficient than $M_1$." Frequently, a quantitative assessment is needed of the extent of $M_0$'s efficiency gain over $M_1$. Some results of this nature will be given in this section and the next.

Empirical and semitheoretical studies of $M$-strategies have been reported by Sampford (1969), Rao and Bayless (1969), Bayless and Rao (1970). The Rao–Bayless studies compare the $p$-unbiased $M$-strategies $M_{HT}$, $M_{HH}$, $M_{RHC}$, $M_{DR}$, and $M_{MU}$ for a broad spectrum of conditions involving seven artificial populations of size $N$ varying from 4 to 6, and some 20 natural populations of sizes varying from 9 to 35. These represented a range of different $x$-to-$y$ correlations, degrees of skewness, and so on. The sample sizes used were $n = 2, 3$, and 4. The "empirical approach" refers to comparison of the $p$-variances computed by evaluation of the estimator for each possible sample. The term "semitheoretical approach" refers to computation of the expected $p$-variances under Model $G_{Rg}$ with $g = 1.0$ (for $n = 2$ only), 1.5, 1.75, and 2.0. Thus in this approach it is pretended that the population value of $y$ for label $k$ has been generated by Model $G_{Rg}$ where $g$ has one of the specified values. The expected variances, given by Theorem 7.1 in the cases of $M_{HT}$, $M_{HH}$, and $M_{RHC}$, are functions of the auxiliary values $x_1, \ldots, x_N$ in the actual finite population, as well as of $g$, $\sigma^2$, $n$ and $N$. Hence, for each population, efficiencies can be computed using the definition (3.2). (It is obviously difficult to ascertain the best fitting value of $g$ in the case of small size populations. The assumption of linear regression through the origin may be given approximate verification by a simple plot of $y$ against $x$.)

Figure 7.1 summarizes the Rao–Bayless findings on expected variances (under Model $G_{Rg}$), and supplements the result of Theorem 7.1. In the figure, a strategy $M_1$ is consistently more efficient in the $\mathscr{E}V$ sense (that is, for *all*

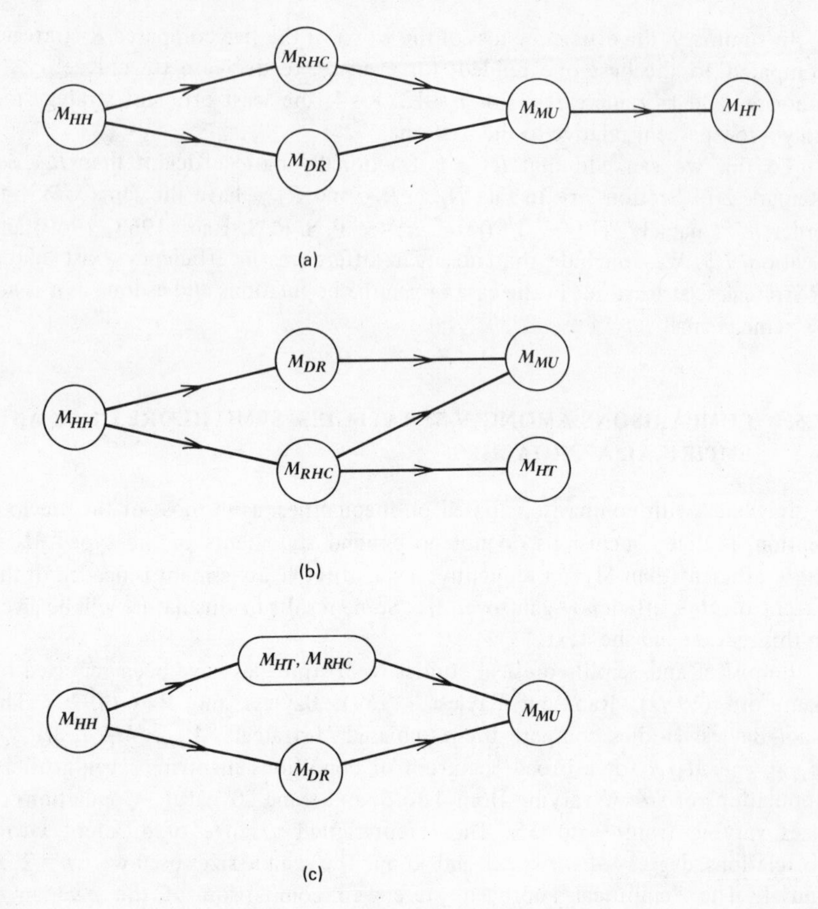


Figure 7.1 Comparison of five $M$-strategies for the Rao–Bayless populations. (a) $g = 2$; (b) $g = 1.5, 1.75$; (c) $g = 1$, in which case $M_{HT}$ and $M_{RHC}$ are equally efficient.

populations in the Rao–Bayless study) than $M_0$, if it is possible to go (always in the direction of the arrows) along a path from $M_0$ to $M_1$.

The following comments provide some further detail on the Rao–Bayless results. Generally speaking, the largest efficiency losses (relative to the best of the investigated $M$-strategies) were observed when $f = n/N$ and $C_x$, the coefficient of variation of $x$, were relatively speaking large.

1. For each of the populations in the study, it was either $M_{HT}$ or $M_{MU}$ that emerged as most efficient among the compared strategies. For $g = 2$, where $M_{HT}$ is uniformly more efficient than $M_{MU}$ by virtue of Theorem 4.2, the efficiency gains of $M_{HT}$ over $M_{MU}$ ranged up to 9%. By contrast, for

$g = 1$, it is known (Section 7.4) that $M_{MU}$ is uniformly more efficient than $M_{HT}$. In fact, $M_{MU}$ was found more efficient than $M_{HT}$ not only when $g = 1$, but also for all investigated populations when $g = 1.5$ and for most of the populations when $g = 1.75$. The efficiency gain of $M_{MU}$ over $M_{HT}$ was usually modest (1 or 2%), but went as high as 12%.

2. For any $\mathbf{x} = (x_1, \ldots, x_N)$, $M_{DR}$ is known to be less efficient than $M_{MU}$ (see Remark 1 of Section 7.3) and also less efficient than $M_{HT}$ if $g = 2$. For a few populations, especially when $g = 1$, $M_{DR}$ actually gained a few percent over $M_{HT}$, but most of the time the relationship was reversed. In one case where $g = 2$, $n = 4$, and $C_x$ was relatively large, $M_{DR}$ lost as much as 30% relative to $M_{HT}$.
3. The strategy $M_{RHC}$ is, by Corollary 7.1, uniformly less efficient than $M_{HT}$ when $g > 1$; for $n = 4$, $g = 2$, and large $C_x$, the efficiency losses relative to $M_{HT}$ went as high as 40%. Also, $M_{RHC}$ turned out to be consistently less efficient than $M_{MU}$ and usually less efficient than $M_{DR}$.
4. The strategy $M_{HH}$ turned out to be consistently less efficient than all other strategies compared. The efficiency loss of $M_{HH}$ relative to $M_{HT}$ tended to be highest for $g = 2$, ranging up to 31% (in a case where $g = 2$, $n = 3, N = 10$) and 57% ($g = 2, n = 4, N = 10$).

In summary, these comparisons of expected variances show that, although $M_{HT}$, $M_{HH}$, and $M_{RHC}$ are equally efficient strategies when $N = \infty$, the efficiency differentials can be substantial if $N$ is small and $f$ is large. As already pointed out, there are disadvantages associated with $M_{HT}$ (intractable design) and $M_{MU}$ (intractable estimator). The practical advantages of $M_{RHC}$ and $M_{HH}$ may, therefore, in certain situations outweigh their efficiency losses.

**Remark 1.** Rao and Bayless (1969), Bayless and Rao (1970) also report the variances (in addition to the *expected* variances discussed above) for the various $M$-strategies and for each of the populations, and $n = 2$, 3, and 4. Lahiri's $R$-strategy, $R_{LA}$, was also included in these comparisons.

Formula (4.3) shows that $V(M_{HT})$ depends on the joint inclusion probabilities $\alpha_{kl}$. (By contrast the *expected* variance (4.4) does not contain the $\alpha_{kl}$, since the $Y_k$ are independent under Model $G_{Rg}$.) Rao and Bayless review various schemes for PPS sampling satisfying $\alpha_k \propto x_k$ ($k = 1, \ldots, N$); such methods may, of course, differ with respect to their associated $\alpha_{kl}$. The Rao-Bayless empirical comparisons show that the selection methods of Brewer (1963a), Fellegi (1963), and Hanurav (1967) mentioned in Section 1.4 give values of $V(M_{HT})$ that differ very little. There was some tendency for Hanurav's (1967) method to be the least efficient of the three.

The impression of relative strength obtained above from comparison of the *expected* variances of $M_{HT}, M_{HH}, M_{RHC}, M_{DR}$, and $M_{MU}$ was, by and large, confirmed by the Rao–Bayless comparisons of the variances:

The Murthy procedure $M_{MU}$ was often (for more than half of the populations) more efficient than either of the three versions of $M_{HT}$, and sometimes $M_{DR}$ and $M_{RHC}$ were also more efficient than $M_{HT}$. The with replacement strategy $M_{HH}$ was inferior to all other $M$-strategies.

The $R$-strategy $R_{LA}$ was more efficient than all of the $M$-strategies for 8 out of 20 natural populations when $n = 2$, but for some of the remaining populations $R_{LA}$ was inferior even to the worst of the $M$-strategies, that is, $M_{HH}$.

Keeping in mind that $R_{LA}$ is essentially the most efficient among the $R$-strategies compared in the previous section, the conclusion is that a choice, based on efficiency alone, between an $R$-strategy and an $M$-strategy is difficult. Available results, including Section 7.6 to follow, indicate that for markedly skewed populations, the situation is volatile: The efficiency of, say, $R_{LA}$ relative to $R_{HT}$ can be either very high or very low. For more symmetric populations, there appears to be less of a difference between the two. (An additional consideration is that the variance estimators reported by Rao and Bayless for $R_{LA}$ are poor.) ■

**Remark 2.** There are at least two well-known methods for estimation of the variance $V(M_{HT})$ of the Horvitz–Thompson strategy, namely, the estimator suggested by Horvitz and Thompson (1952),

$$\hat{V}_{HT} = \Sigma_S(1 - \alpha_k)\left(\frac{y_k}{\alpha_k}\right)^2 + \Sigma\Sigma\left(\frac{\alpha_{kl}}{\alpha_k\alpha_l} - 1\right)\left(\frac{y_k y_l}{\alpha_k\alpha_l}\right)$$

and the one suggested by Yates and Grundy (1953) and Sen (1953),

$$\hat{V}_{YG} = \Sigma\Sigma\left(\frac{\alpha_k\alpha_l}{\alpha_{kl}} - 1\right)\left(\frac{y_k}{\alpha_k} - \frac{y_l}{\alpha_l}\right)^2$$

where $\Sigma\Sigma$ is over $k \in S$, $l \in S$, $k \neq l$. Provided $\alpha_{kl} > 0$ for all $k$ and $l$, the estimator $\hat{V}_{HT}$ is $p$-unbiased for $V(H_{MT})$. A sufficient condition for $\hat{V}_{YG}$ to be $p$-unbiased for $V(H_{MT})$ is that $p$ is an FES($n$) design and that $\alpha_{kl} > 0$ for all $k$ and $l$.

An important practical question concerns the stability of a variance estimator, that is, its tendency to show relatively little sample fluctuations. A minimum requirement of stability is that the variance estimator be always nonnegative. The estimator $\hat{V}_{YG}$ is usually preferred over $\hat{V}_{HT}$ on the grounds that the former is always nonnegative in certain cases where the latter may not be. A sufficient condition for $\hat{V}_{YG}$ to be always nonnegative is that $\alpha_{kl} < \alpha_k\alpha_l$ for all $k$, $l$. For certain designs, $\hat{V}_{YG}$ has been shown to be always nonnegative; see Sen (1953), Raj (1956), J. N. K. Rao (1961, 1963a, 1965), Seth (1966), Lanke (1975).

Hanurav (1967) has suggested that one might require that min $\alpha_{kl}/\alpha_k\alpha_l > c$, for $c$ sufficiently away from zero, as a means of improving the stability of $\hat{V}_{YG}$. He also gave a method for PPS selection satisfying the requirement.

Rao and Bayless (1969), Bayless and Rao (1970) also studied the stability of customary unbiased variance estimators for $M_{HT}$, $M_{HH}$, $M_{RHC}$, $M_{DR}$, $M_{MU}$, and $R_{LA}$. The results show that the variance estimators of $M_{RHC}$, $M_{DR}$, and $M_{MU}$ tend to be more stable than the variance estimators of $M_{HT}$ for the various methods of selection. The unbiased variance estimator of $R_{LA}$ given by Raj (1954) and Sen (1955) tended to be the least stable among the strategies studied. ■

## 7.6. COMPARISONS BASED ON THE GAMMA($r$) APPROXIMATION

In summary of Sections 7.4–7.5, we found that the efficiency differences between the various $M$-strategies tended to disappear as the population size $N$ gets larger (although, for $M_{MU}$ the picture is not quite clear). Likewise, the efficiency differences within the set of $R$-strategies were relatively minor for large $N$ and if $n$ is not too small. For moderately large to very large populations, we can thus treat $M$-strategies as one group and $R$-strategies as a competing group and ask the question: Should an $M$-strategy be preferred to an $R$-strategy, or vice versa?

Put differently, do $p$-unbiased strategies based on PPS sampling ever pay off, considering that almost all the $R$-strategies enjoy the advantages of simple random sampling, while the $M$-strategies suffer from the various disadvantages of PPS sampling referred to earlier? As we shall see, there are conditions under which large gains in efficiency are indeed realizable with an $M$-strategy.

In order to facilitate assessment, especially of the mathematically cumbersome $R$-strategies, assume that $N$ is very large and that the frequency distribution of the auxiliary variable values $x_1, \ldots, x_N$ is approximated by a continuous curve of known shape. Mathematical convenience, as well as a broad range of shapes, is achieved if we assume that distribution of the $x$-values in the population is described by the gamma($r$) density

$$f_r(x) = \begin{cases} x^{r-1} e^{-x}/\Gamma(r) & \text{for } x > 0 \\ 0 & \text{for } x \leqslant 0 \end{cases} \tag{6.1}$$

Assume that identification of a unit is made on the basis of its $x$-value rather than by its label. Thus one single draw from the population is pictured as the selection of the value $x$ with a certain specified probability density not necessarily equal to $f_r(x)$, since unequal probability sampling is permitted. The associated value of the variable of interest $y = y_x$ is then observed.

For example, in one completely random draw, the value $x$ is selected with probability density $p(x) = f_r(x)$ for $x > 0$ and $p(x) = 0$ elsewhere. On the other hand, in one draw according to PPS of $x$, selection of the value $x$ is executed according to the probability density $p(x) = xf_r(x)/r$ for $x > 0$ and $p(x) = 0$ elsewhere. The analogy with PPS sampling in the finite setting is that the selection probabilities $p_k = x_k/N\bar{x}$ such that $\Sigma_1^N p_k = 1$ correspond in the infinite setting to the selection density $p(x) = xf_r(x)/r$ such that $\int_0^\infty p(x)\,dx = 1$, noting that $r$ is the mean of $x$ in the gamma($r$) population.

The following definition of the gamma($r$) approximation technique is based on Cassel and Särndal (1972, 1974).

Approximative evaluation of the $\mathscr{E}$MSE of a strategy according to the *gamma(r) approximation* will be based on the following assumptions:

(i) $N = \infty$;

(ii) If $x$ denotes the auxiliary variable value associated with the $i$th of $n$ draws, a sample of $n$ $x$-values, $\mathbf{x} = (x_1, \ldots, x_n)$, is selected in accordance with the specified function $p(\mathbf{x})$ satisfying

$$\int_{R_n} p(\mathbf{x})\,d\mathbf{x} = 1$$

(iii) If $n$ simple random sampling draws were to be made, the probability density of $\mathbf{x}$ would be $p(\mathbf{x}) = f(\mathbf{x})$, where

$$f(\mathbf{x}) = \prod_{i=1}^{n} f_r(x_i)$$

and $f_r(x)$ is given by (6.1).

The choice of a "design function" $p(\mathbf{x})$ corresponds to the choice of a sampling design $p(s)$ in the finite setting. For a given function $p(\mathbf{x})$ satisfying (ii), the ratio $p(\mathbf{x})/f(\mathbf{x})$ represents the probability density of the outcome $\mathbf{x}$ relative to the frequency with which $\mathbf{x}$ would occur in $n$ simple random sampling draws (with or without replacement). From (ii) it follows that

$$\int_{R_n} \left\{\frac{p(\mathbf{x})}{f(\mathbf{x})}\right\} f(\mathbf{x})\,d\mathbf{x} = 1$$

In this approach, no difference is made between with and without replacement sampling, because of the infinite population size. To the designs denoted *srs* and *srsr* in the finite setting corresponds the density $p(\mathbf{x})$ such that, for all $\mathbf{x} > \mathbf{0}$,

$$\frac{p(\mathbf{x})}{f(\mathbf{x})} = 1$$

By Theorem 7.1, the strategies $M_{HT}$, $M_{RHC}$, and $M_{HH}$ are equally efficient when $N$ is infinite. To their respective designs, *ppsx*, *ppsgx*, and *ppsrx*, corresponds $p(\mathbf{x})$ such that, for $x > \mathbf{0}$,

$$\frac{p(\mathbf{x})}{f(\mathbf{x})} = \prod_{i=1}^{n} \left(\frac{x_i}{r}\right) \tag{6.2}$$

To the design $pps\Sigma x$ corresponds $p(\mathbf{x})$ such that, for $\mathbf{x} > \mathbf{0}$,

$$\frac{p(\mathbf{x})}{f(\mathbf{x})} = \sum_{i=1}^{n} \frac{x_i}{nr} \tag{6.3}$$

This is seen by noting that in the finite setting, formula (3.1) gives $p(s) = \bar{x}_s / \binom{N}{n} \bar{x}$ under the design $pps\Sigma x$, which should be set in relation to $p(s) = 1/\binom{N}{n}$ under the design *srs*.

**Remark 1.** Our family of gamma populations is not as restricted as it looks. With equivalent results, we could have assumed that the population is such that $x/\gamma$ is gamma($r$) distributed, where $\gamma$ is any positive scale parameter. Nothing in the comparison of two strategies would be changed: The ratio of two $\mathscr{E}$MSE's remains unaltered. ■

As a result of $n$ draws, we obtain the observations $(x_i, y_i)$ $(i = 1, \ldots, n)$. The value $y_i$ is assumed to be the outcome of the random variable $Y_i$ $(i = 1, \ldots, n)$, for which Model $G_{Rg}$ will be assumed. In this setting, the model will imply that, for $i = 1, \ldots, n$, the variables $Y_1, \ldots, Y_N$ are independent, and

$$\begin{aligned} &\mathscr{E}(Y_i \mid x_i) = \beta x_i;\ \mathscr{E}\{(Y_i - \beta x_i)^2 \mid x_i\} = \sigma^2 x_i^g \\ &\mathscr{E}\{(Y_i - \beta x_i)(Y_j - \beta x_j) \mid x_i, x_j\} = 0 \quad (i \neq j) \end{aligned} \tag{6.4}$$

The ratio estimator is $t_R = \bar{x}(\Sigma_1^n y_i)/(\Sigma_1^n x_i)$, the estimator $\bar{x}R_{yx}$ becomes $\bar{x}\,\Sigma_1^n y_i/nx_i$.

The following theorem expands on the comparison of $R_{LA}$ with $M_{HT}$ in Theorem 7.2:

**Theorem 7.3.** *Under the formulation (6.4) of Model $G_{Rg}$, and under the gamma($r$) approximation, the following holds, provided $r + g - 1 > 0$:*

$$\mathscr{E}V(M_{HT}) = \frac{\{r\Gamma(r+g-1)/\Gamma(r)\}\sigma^2}{n}$$

$$\mathscr{E}V(R_{LA}) = \frac{\{\Gamma(r+g)/\Gamma(r)\}\{1 + (g-1)/nr\}^{-1}\sigma^2}{n}$$

*Hence the efficiency gain (or loss) of $M_{HT}$ relative to $R_{LA}$, as defined by (3.2), is given by*

$$A(M_{HT}; R_{LA}) = \frac{(n-1)(g-1)}{nr + g - 1} \tag{6.5}$$

**Proof.** Both $M_{HT}$ and $R_{LA}$ are $p$-unbiased as well as $\xi$-unbiased. Therefore, by Lemma 4.1, formula (3.4), $\mathscr{E}V(M_{HT}) = E\,\mathscr{V}(\bar{x}R_{Yx}) - \mathscr{V}(\bar{Y})$, where $E(\cdot)$ is with respect to the design $ppsx$. Similarly, $\mathscr{E}\,V(R_{LA}) = E\,\mathscr{V}(T_R) - \mathscr{V}(\bar{Y})$, where $E(\cdot)$ is with respect to the design $pps\Sigma x$. Under Model $G_{Rg}$, $\mathscr{V}(\bar{Y}) = m_g\sigma^2/N$, where $m_g$ is given by (4.2). Since the gamma($r$) approximation implies $N = \infty$, $\mathscr{V}(\bar{Y})$ vanishes.

It remains to evaluate the term $E\,\mathscr{V}(\cdot)$ for each of the two strategies. For $M_{HT}$ we have

$$E\,\mathscr{V}(M_{HT}) = \frac{\bar{x}^2\sigma^2 E(\Sigma_1^n x_i^{g-2})}{n^2}$$

where $\bar{x} = r$ and $E(\cdot)$ is with respect to the density $p(\mathbf{x}) = \Pi_1^n\{x_i^r e^{-x_i}/\Gamma(r+1)\}$ obtained from (6.2). Hence, $E(\Sigma_1^n x_i^{g-2}) = n\Gamma(r+g-1)/\Gamma(r+1)$, which gives the expression for $\mathscr{E}\,V(M_{HT})$ stated in the theorem.

Turning to $R_{LA}$, we have after simplification

$$E\,\mathscr{V}\,(R_{LA}) = \bar{x}^2\sigma^2 E\left\{\frac{\Sigma_1^n x_i^g}{(\Sigma_1^n x_i)^2}\right\}$$

where $E(\cdot)$ is with respect to the density

$$p(\mathbf{x}) = \frac{\Sigma_1^n x_i}{nr}\Pi_1^n\{x_i^{r-1}e^{-x_i}/\Gamma(r)\}$$

obtained from (6.3). Proceeding as in Rao and Webster (1966),

$$E\left\{\frac{\Sigma_1^n x_i^g}{(\Sigma_1^n x_i)^2}\right\} = \frac{\Sigma_1^n\{\Gamma(r+g)/\Gamma(r)\}E_{oi}(1/\Sigma_1^n x_i)}{nr}$$

where $E_{oi}(\cdot)$ is with respect to the $n$-dimensional density of independent random variables $X_1, \ldots, X_n$, where $X_i$ has the gamma($r+g$) density and the remaining $(n-1)$ variables all have the gamma($r$) density. Under these conditions, $\Sigma_1^n X_i$ has the gamma($nr+g$) density; thus $E_{oi}(1/\Sigma_1^n X_i) = 1/(nr+g-1)$, for $i = 1, \ldots, n$, which gives $\mathscr{E}\,V(R_{LA})$ as stated in the theorem. □

**Remark 2.** Using the gamma($r$) approximation as in Theorem 7.3, it is easily shown that, under Model $G_{Rg}$,

$$\mathscr{E}\text{MSE}(R_C) = \left\{\frac{\Gamma(r+g)}{\Gamma(r)}\right\}\left\{\frac{nr^2}{(nr+g-1)(nr+g-2)}\right\}\sigma^2$$

From Theorem 7.3 and formula (3.2), the efficiency gain (loss) of $R_C$ relative to $R_{LA}$ is computed as $A(R_C; R_{LA}) = (g-2)/nr$. Hence, if $0 \leqslant g < 2$ and unless $n$ and $r$ are extremely small, the $p$-unbiased Lahiri strategy $R_{LA}$ is just marginally better, in the $\mathscr{E}$MSE sense, than the classical $R$-strategy $R_C$. If $g = 2$, the two strategies are equally efficient. The $\mathscr{E}$MSE's of $R_C$ and $R_{LA}$ are retrievable from P. S. R. S. Rao (1968). ■

Based on Theorem 7.3, formula (6.5), Table 7.2 shows the percentage efficiency gain (loss) of $M_{HT}$ relative to $R_{LA}$ for selected values $g$, $n$, and $r$. For $g = 1$, the two strategies are equally efficient.

**Remark 3.** Table 7.2 reveals the considerable efficiency difference that may exist, for skewed populations and if $g \neq 1$, between $M_{HT}$ and $R_{LA}$, and hence also in the comparison between any $M$-strategy and any $R$-strategy, among the ones discussed. By Table 7.2, if $g > 1$, and particularly for skewed populations ($r$ small), $M_{HT}$ is greatly superior to $R_{LA}$. On the other hand, if $g < 1$, $M_{HT}$ can be an extremely inefficient procedure compared to $R_{LA}$, especially if the population is skewed.

In practice, $g$ is usually unknown in Model $G_{Rg}$, but can perhaps be specified to within an interval. Table 7.2 suggests the following rules of thumb applicable when $N$ is large, $n$ is not extremely small, and the population is approximately gamma-shaped:

If $1 \leqslant g \leqslant 2$, an $M$-strategy (for example, $M_{HT}$ or $M_{RHC}$) appears to be a good choice. In particular, it is strongly preferred to its $R$-strategy competitors if the population is highly skewed.

If $0 \leqslant g \leqslant 1$, $M_{HT}$ (and $M_{RHC}$) should be avoided, especially if the population is skewed, and $R_{LA}$ or $R_C$ should be used. (If $r + g \leqslant 1$, $\mathscr{E}\text{MSE}(M_{HT})$ is in fact infinite.)

If $0 \leqslant g \leqslant 2$, an $R$-strategy seems to be a more prudent approach than an $M$-strategy. Put differently, if the population is highly skewed, the choice of an $R$-strategy would at least avoid a highly inefficient procedure.

**Table 7.2. Percentage Efficiency Gain (Loss), 100 $A(M_{HT};R_{LA})$, under Model $G_{Rg}$ and the Gamma($r$) approximation. For $g = 1$ the strategies are equally efficient.**

| | $g = 0$ | | $g = 1.5$ | | $g = 2$ | |
|---|---|---|---|---|---|---|
| $n$ | $r = 1$ | $r = 4$ | $r = 1$ | $r = 4$ | $r = 1$ | $r = 4$ |
| 2 | −100 | −14 | 20 | 6 | 33 | 11 |
| 4 | −100 | −20 | 33 | 9 | 60 | 18 |
| 10 | −100 | −23 | 43 | 11 | 82 | 22 |
| 20 | −100 | −24 | 46 | 12 | 90 | 23 |
| $\infty$ | −100 | −25 | 50 | 13 | 100 | 25 |

If $g$ is approximately 0, it is possible that both $M_{HT}$ and $R_{LA}$ can be considerably improved upon by some alternative strategy, but as yet no such strategy seems available.

If the gamma-shaped population has a negligible skewness, then it is a tossup between an $M$-strategy and an $R$-strategy from an efficiency point of view, and the $R$-strategy may be preferred because of its simple design. ■

This concludes our case study of $M$-strategies and $R$-strategies. Our interest in these strategies does not necessarily imply that they make better use of auxiliary information in all situations than, say, other types of strategy mentioned in Table 7.1; in fact, Remark 4 below is an example showing that improvements are possible. In particular, the picture remains incomplete in regard to the possibilities of the generalized ratio estimator and the generalized regression estimator contained in Table 7.1.

**Remark 4.** Under Model $G_{Rg}$, the generalized regression estimator (see Remark 6 of Section 5.4, and Table 7.1) is given by

$$t_{GREog} = m_{t1} R_{yt} + \hat{\beta}(\bar{x} - m_{t1} R_{xt})$$

where $t_k = x_k^{g/2}$, $m_{t1} = \Sigma_1^N t_k/N$, $R_{yt} = \Sigma_S y_k/nt_k$, $R_{xt} = \Sigma_S x_k/nt_k$, and $\hat{\beta} = \{\Sigma_S x_k y_k/t_k^2\}/\{\Sigma_S x_k^2/t_k^2\}$.

Let $ppsx^{g/2}$ denote an FES($n$) design having the inclusion probabilities $\alpha_k \propto x_k^{g/2}$ $(k = 1, \ldots, N)$, assuming $nx_k^{g/2} < Nm_{t1}$ for all $k$. The $\xi$-unbiased strategy

$$H_{GRg} = (ppsx^{g/2}, t_{GREog})$$

was suggested by Cassel, Särndal, and Wretman (1976).

If $g = 2$, $H_{GRg}$ is identical to the $p$-unbiased strategy $M_{HT}$. If $g = 0$, $H_{GRg}$ becomes $(p_o, t_{REG})$ where $p_o$ is any FES($n$) design with constant inclusion probabilities $\alpha_k = f = n/N$ for $k = 1, \ldots, N$, and $t_{REG} = \bar{y}_S + \hat{\beta}(\bar{x} - \bar{x}_S)$, the approximately $p$-unbiased regression estimator.

If the value of $g$ is such that $0 < g < 2$, then the strategy $H_{GRg}$ utilizes the auxiliary information $(x_1, \ldots, x_N)$ at *both* the design stage and the estimation stage, and the $p$-bias of $H_{GRg}$ is small if the $(x_k, y_k)$-scatter indicates a linear relation through the origin.

Let us further examine the case $g = 1$. Theorem 7.3 shows that $R_{LA}$ and $M_{HT}$ are equally efficient $p$-unbiased strategies under Model $G_{R1}$ and for the gamma($r$) approximation. Under the same assumptions, the (somewhat $p$-biased) procedure $H_{GR1}$ often has considerably smaller $\mathscr{E}$MSE than either $R_{LA}$ or $M_{HT}$, except if $n$ is extremely small, as shown in Cassel, Särndal, and Wretman (1976). The efficiency gain of $H_{GR1}$ over $R_{LA}$ or $M_{HT}$ increases with $n$ and is especially pronounced if the population is skewed: For $n \geqslant 10$, the gain is at least 47, 22, and 10% for $r = 1/2$, 1, and 2, respectively.

Not surprisingly, it may also be shown that $(srs, \bar{Y}_s)$, which completely ignores the auxiliary information, is a very inefficient alternative, under Model $G_{Rg}$, to either of $H_{GRg}$, $R_{LA}$, and $M_{HT}$ as soon as the correlation between $y$ and $x$ is reasonably high (on the average over the populations generated under the model). ■

## 7.7. ROBUSTNESS PROBLEMS IN MODEL-BASED INFERENCE

In Sections 7.4–7.6 the focus was on studying (approximately) $p$-unbiased strategies from a robustness point of view. More specifically, our case study compared $M$- and $R$-strategies under the various conditions expressed by shifts in the value of $g$ in the selected Model $G_{Rg}$, and by differences in the shape of the distribution of the auxiliary variable. Now we shall study robustness of predictors derived by the model-based approach of Chapter 5, that is, inferences in which little or no attention is paid to design-associated aspects such as $p$-unbiasedness.

Two examples, Cases A and B below, illustrate the type of robustness questions that may arise in model based inference. In either case, the problem arises because the inference is based on an *assumed* model which is not identical to the *true* one.

*Case A. Misspecification of the Variance Function, Model $G_R$.* The Brewer–Royall predictor discussed in Section 5.4,

$$T_{BR} = f_s\bar{Y}_s + (1 - f_s)\bar{x}_{\bar{s}}\hat{\beta} \tag{7.1}$$

with

$$\hat{\beta} = \frac{\Sigma_s x_k Y_k / v(x_k)}{\Sigma_s x_k^2 / v(x_k)}$$

is the $\xi$-BLU predictor of $\bar{Y}$ under Model $G_R$. The form of $T_{BR}$ depends on the specification of the function $v(x)$. As before Model $G_{Rg}$ refers to the special case where $v(x) = x^g$, and the predictor (7.1) is then denoted $T_{BRg}$.

If Model $G_{Rg_0}$ has been assumed, then $T_{BRg_0}$ is the predictor believed to be optimal. If in reality the alternative Model $G_{Rg_1}$ is true, then $T_{BRg_0}$ is still $\xi$-unbiased but less efficient than $T_{BRg_1}$, which is the preferred estimator. That is, $T_{BRg_1}$ is $\xi$-BLU under the true model and should have been chosen, had we known that $g = g_1$.

Thus in this case the only misspecification of the model is in the variance function $v(x)$, and this will result in a certain loss of efficiency. The assumed linear regression through the origin is not in question.

Now, for fixed values $g_0$ and $g_1$, neither of which is necessarily true, let us compare the predictors $T_{BRg_0}$ and $T_{BRg_1}$, both of which are $\xi$-unbiased under any specification of $v(x)$ in Model $G_R$. The following statement, given without proof and due to Royall (1970a), shows that a preference for one or the other predictor can be stated under fairly general conditions on the true variance function $v(x)$.

**Theorem 7.4.** *If $0 \leqslant g_0 \leqslant g_1$, then, for any FES(n) design p, and for any specification of the function $v(x)$ in Model $G_R$ such that $v(x)/x^{g_0}$ is nonincreasing,*

$$\mathscr{E}\text{MSE}(p, T_{BRg_0}) \leqslant \mathscr{E}\text{MSE}(p, T_{BRg_1}) \tag{7.2}$$

*For any function $v(x)$ such that $v(x)/x^{g_1}$ is nondecreasing, the inequality in (7.2) is reversed. For strict inequality in (7.2), it is sufficient that $p(s) > 0$ for some s such that $\nu(s) = n$ and that $x_k \neq x_l$ for some $k \neq l$ in s.* □

For example, if $v(x) = x^2$, we have

$$\mathscr{E}\text{MSE}(p, T_{BR2}) \leqslant \mathscr{E}\text{MSE}(p, T_{BR1}) \leqslant \mathscr{E}\text{MSE}(p, T_{BR0}),$$

where in particular $T_{BR1} = T_R = \bar{x}\bar{Y}_s/\bar{x}_s$, the ratio predictor. If $v(x) = 1$, both inequalities are reversed.

Thus in a comparison of $T_{BR1} = T_R$ with $T_{BR2}$, Theorem 7.4 says that $T_R$ is at least as good as $T_{BR2}$ from an $\mathscr{E}$MSE point of view for any Model $G_{Rg}$ with $g \leqslant 1$, while the reverse is true if $g \geqslant 2$. If $v(x) = x^g$ with $1 \leqslant g \leqslant 2$, the theorem gives no information as to whether $T_R$ or $T_{BR2}$ is the preferred predictor. The design and the particular configuration of $x$-values would decide the comparison. Empirical evaluations may then have to be used to designate the preferred predictor.

*Case B. Misspecification in a Polynomial Regression Model.* Consider a special case of Model $G_{MR}$ (see Table 4.1), namely, the polynomial regression model such that, for $k = 1, \ldots, N$, the variables $Y_1, \ldots, Y_N$ are independent, and

$$\mathscr{E}(Y_k) = \sum_{j=0}^{J} \delta_j \beta_j x_k^j$$

$$\mathscr{V}(Y_k) = \sigma^2 v(x_k), \quad \mathscr{C}(Y_k, Y_l) = 0 \quad (k \neq l)$$

where $\delta_j$ is 0 or 1 according as the term $x_k^j$ is absent or present, respectively, in the model, and $v(\cdot)$ is a known function. Following Royall and Herson (1973a) we denote this model $\xi(\delta_0, \delta_1, \ldots, \delta_J : v(x))$. Thus Model $G_R$ can be described as $\xi(0, 1: v(x))$.

Theorem 7.5 gives the $\xi$-BLU predictor of $\bar{Y}$, in this situation, assuming that all $\beta_j$ are estimable. From a robustness point of view, the problem with the resulting predictor is that the property of $\xi$-unbiasedness is tied to a particular configuration $\delta_0, \delta_1, \ldots, \delta_J$. If there is a *true* model characterized by a certain configuration $\delta_0, \delta_1, \ldots, \delta_J$ and an *assumed* model for which the configuration is different, then the $\xi$-BLU estimator obtained from Theorem 7.5 for the assumed model is usually $\xi$-biased under the true model. We shall discuss a resolution of this problem suggested by Royall and Herson (1973a,b).

**Theorem 7.5.** *Under the model $\xi(\delta_0, \delta_1, \ldots, \delta_J : v(x))$, and for known auxiliary variable measurements $x_k > 0$ $(k = 1, \ldots, N)$, the $\xi$-BLU predictor of $\bar{Y}$ is, for any design p, given by*

$$T = f_s \bar{Y}_s + (1 - f_s) \sum_{j=0}^{J} \delta_j m_{\bar{s}j} \hat{\beta}_j \tag{7.3}$$

*where, for $j = 0, 1, \ldots, J$,*

$$m_{\bar{s}j} = \frac{\Sigma_{\bar{s}} x_k^j}{N - v(s)}$$

*and the $\hat{\beta}_j$ are the least squares estimates of the $\beta_j$ under the model $\xi(\delta_0, \delta_1, \ldots, \delta_J : v(x))$.*

**Proof.** The result follows easily from Theorem 5.3 noting that the desired predictor must be of the form $f_s \bar{Y}_s + (1 - f_s)U$, where $\mathscr{E}(U) = \mu_{\bar{s}} = \Sigma_{j=0}^{J} \delta_j m_{\bar{s}j} \beta_j$. □

Royall and Herson's approach, illustrated below for the case where $J = 1$ and $\xi(0, 1 : x)$ is the assumed model, will preserve $\xi$-unbiasedness under a variety of polynomial regression models:

Since $\xi(0, 1 : x)$ is assumed, (7.1) says that the predictor believed to be $\xi$-BLU is $T_{BR1} = T_R = \bar{x}\bar{Y}_s/\bar{x}_s$, the classical ratio predictor. We also know that, under the conditions of Theorem 5.5, the $\mathscr{E}$MSE is minimized by the strategy $(p^*, T_R)$, where $p^*$ is the purposive design (4.8) of Section 5.4, which selects, with probability one, the set of labels corresponding to the $n$ largest $x$-values.

If, however, the alternative model $\xi(\delta_0, \delta_1, \ldots, \delta_J : v(x))$ is really true, then the preferred predictor, regardless of design, is given by (7.3). Moreover, $T_R$ is $\xi$-biased under $\xi(\delta_0, \delta_1, \ldots, \delta_J : v(x)) \neq \xi(0, 1 : v(x))$. The approach suggested by Royall and Herson (1973a,b) attempts to "save" the inference $T_R$ by the only avenue that remains open, namely, to choose the design so as to produce only a limited set of samples $s$ such that $T_R$ remains $\xi$-unbiased under more general models than those of type $\xi(0, 1 : v(x))$. The following argument leads to the notion of a balanced sampling design:

If the *true* model is $\xi(\delta_0, \delta_1, \ldots, \delta_J : v(x))$, the $\xi$-bias of $T_R = \bar{x}\bar{Y}_s/\bar{x}_s$ is

$$\mathscr{E}(T_R - \bar{Y}) = \sum_{j=0}^{J} \delta_j \beta_j m_1 \left\{ \frac{m_{sj}}{m_{s1}} - \frac{m_j}{m_1} \right\}$$

where

$$m_{sj} = \frac{\Sigma_s x_k^j}{v(s)}$$

and $m_j$ is defined by (4.2). The $\xi$-bias is zero if and only if

$$\frac{m_{sj}}{m_{s1}} = \frac{m_j}{m_1} \tag{7.4}$$

for all $j$ such that $\delta_j = 1$ in the model $\xi(\delta_0, \delta_1, \ldots, \delta_J : v(x))$. This leads to the following definition of a balanced sample.

**Definition.** A *balanced sample*, denoted $s(J)$, is any sample satisfying (7.4) for $j = 0, 1, \ldots, J$, that is, $s(J)$ is such that

$$m_{sj} = m_j \tag{7.5}$$

for $j = 1, 2, \ldots, J$. A sample $s$ such that (7.5) holds for $j = j_0 \leqslant J$ is said to be *balanced on the* $j_0$th *moment*. A design $p$ which selects, with probability one, a balanced sample will be called a *balanced (sampling) design* and will be denoted $pba_J$. ■

**Remark 1.** A population is completely specified by its moments. Since a balanced sample has its first $J$ moments equal to those of the population, such a sample evokes the notion of being "representative" for the population. The actual selection of a sample that is balanced on each of $J$ moments poses obvious practical difficulties. Royall and Herson (1973a) admit this difficulty and advise that "simple random sampling will frequently produce a sample which is a fair approximation to $s(J)$ for some $J \geqslant 1$". ■

**Remark 2.** Under a balanced sampling design we have $T_R = \bar{Y}_s$, the sample mean. However, the *strategy* $R_{BAL} = (pba_J, T_R) = (pba_J, \bar{Y}_s)$, which is in general $p$-biased, must not be confused with the traditional $p$-unbiased sample mean strategy $(srs, \bar{Y}_s)$. ■

Selection of a sample that is balanced on the first moment guarantees the protection of $T_R = \bar{Y}_s$ against the $\xi$-bias that would arise if $\xi(1, 1 : v(x))$ were the true model. The price of this protection is a certain loss of efficiency. Assuming FES($n$) designs, we compare the balanced sampling strategy $R_{BAL} = (pba_1, T_R)$ to $R_{OPT} = (p^*, T_R)$, which has minimum $\mathscr{E}$MSE under $\xi(0, 1 : x)$, $p^*$ being given by (4.8) of Section 5.4.

Using (5.1) of Section 5.5 we have

$$\mathscr{E}\,\text{MSE}(R_{OPT}) = \min_{s \in \mathscr{S}_n} \left(\frac{\bar{x}_{\bar{s}}}{\bar{x}_s}\right) \frac{(1-f)\bar{x}\sigma^2}{n}$$

$$\mathscr{E}\,\text{MSE}(R_{BAL}) = \frac{(1-f)\bar{x}\sigma^2}{n}$$

where $\mathscr{S}_n = \{s : \nu(s) = n\}$. The efficiency loss is the absolute value of

$$A(R_{BAL}; R_{OPT}) = \min_{s \in \mathscr{S}_n} \left( \frac{\bar{x}_{\bar{s}}}{\bar{x}_s} \right) - 1 \leqslant 0 \tag{7.6}$$

Royall and Herson (1973a) calculated the value taken by (7.6) for $N = \infty$; $f = 1/2, 1/8, 0$; and three population shapes (uniform, symmetric triangular, and right triangular with negative slope), each distributed on the interval $(a, \gamma a)$, for arbitrary $a$.

The general conclusion was that shape is a less important factor in determining the efficiency loss than is $\gamma$, the ratio of the extremes of the distribution. The greatest loss, 74%, was obtained for $\gamma = \infty$, $f = 1/2$, right triangular shape; in many other cases the loss exceeded 40%. Hence, the protection against $\xi$-bias is often costly from an efficiency point of view.

**Remark 3.** The balanced design $pba_1$ protects the predictor $T_R$ against the $\xi$-bias which would be incurred if $\xi(1, 1 : v(x))$, not $\xi(0, 1 : x)$, is the true model. An attractive property of balanced sampling designs is that if the conditions $m_{sj} = m_j$ $(j = 2, 3, \ldots, J)$ are also satisfied, then $T_R$ is protected against $\xi$-bias under any model $\xi(\delta_0, \delta_1, \ldots, \delta_J : v(x))$, yet there is no additional loss of efficiency; that is, (7.6) also represents the loss of efficiency of $(pba_J, T_R)$ relative to $R_{OPT}$. ■

**Remark 4.** Under Model $G_{R1}$, we have $\mathscr{E}\text{MSE}(R_{BAL}) = \mathscr{E}V(R_{LA}) = (1 - f)\bar{x}\sigma^2/n$, that is, $R_{BAL} = (pba_1, T_R)$ and $R_{LA} = (pps\Sigma x, T_R)$ are equally efficient strategies. This follows from (5.1) of Section 5.5 since $E(\bar{x}_{\bar{s}}/\bar{x}_s) = 1$ for both designs, $pba_1$ and $pps\Sigma x$. In the latter case, use of (3.1) and (4.5) helps to establish the result.

Moreover, $R_{LA}$ and the classical ratio procedure $R_C$ are often just about equally efficient; see Remark 2 of Section 7.6.

Thus we conclude that for the model $\xi(0, 1 : x)$, balancing the design is on the average equivalent to $p$-unbiasedness, and that the protection sought by balancing (see Remark 1 above) wipes out the efficiency gain realizable under the model based approach if we were willing to accept an extreme, purposive sample as a basis for the inference.

On the other hand, Remark 4 of Section 7.6 showed that efficiency can be gained by appropriate use of the auxiliary information at the design stage: The strategy $H_{GRg}$ for $g = 1$ is often considerably more efficient under Model $G_{R1}$ than either of three equally efficient competitors: $R_{BAL}$, $R_{LA}$, and $M_{HT}$. ■

**Remark 5.** It was observed that $T_R$ reduces to $\bar{Y}_s$ under balanced sampling. More generally, balanced sampling has the property of reducing the $\xi$-BLU predictor of $\bar{Y}$ to $\bar{Y}_s$: If $T = T(\delta_0, \delta_1, \ldots, \delta_J : v(x))$ denotes the $\xi$-BLU predictor (7.3) given in Theorem 7.5, then, under the balanced design $pba_J$,

$$T(1, \delta_1, \delta_2, \ldots, \delta_J : 1) = T(\delta_0, 1, \delta_2, \ldots, \delta_J : x) = T(\delta_0, \delta_1, \delta_2, \ldots, 1 : x^J) = \bar{Y}_S$$

for any configuration $\delta_0, \delta_1, \ldots, \delta_J$ of 0's and 1's. The result is due to Royall and Herson (1973a). ∎

**Remark 6.** In the preceding discussion, the idea of balanced sampling grew out of an example involving the ratio estimator, but obviously the concept has wider applicability.

Under the model $\xi(1, 1 : 1)$ we showed in Example 2 of Section 5.6 that the $\xi$-BLU predictor is the classical regression predictor

$$T_{REG} = \bar{Y}_s + \hat{\beta}(\bar{x} - \bar{x}_s)$$

with

$$\hat{\beta} = \frac{\Sigma_s(x_k - \bar{x}_s)Y_k}{\Sigma_s(x_k - \bar{x}_s)^2}$$

Now if the alternative model $\xi(\delta_0, \delta_1, \ldots, \delta_J : v(x))$ were actually true, it is easily verified that $T_{REG}$ is in general $\xi$-biased, and that the $\xi$-bias can be removed if we choose the balanced design $pba_J$. For any balanced sample, the predictor $T_{REG}$ reduces to the sample mean $\bar{Y}_s$. ∎

# References


**Aggarwal, O. P.** (1959). Bayes and minimax procedures in sampling from finite and infinite populations I. *Ann. Math. Statist.* **30** 206–218.

**Aggarwal, O. P.** (1966). Bayes and minimax procedures for estimating the arithmetic mean of a population with two-stage sampling. *Ann. Math. Statist.* **37** 1186–1195.

**Avadhani, M. S. and B. V. Sukhatme** (1970). A comparison of two sampling procedures with an application to successive sampling. *Appl. Statist.* **19** 251–259.

**Asok, C. and B. V. Sukhatme** (1976). On Sampford's procedure of unequal probability sampling without replacement. *J. Amer. Statist. Ass.* **71** 912–918.

**Barnard, G. A.** (1963). The logic of least squares. *J. R. Statist. Soc. B* **25** 124–127.

**Barnard, G. A.** (1971). Discussion of paper by V. P. Godambe and M. E. Thompson. *J. R. Statist. Soc. B* **33** 376–378.

**Barnard, G. A, G. M. Jenkins, and C. B. Winsten** (1962). Likelihood inference and time series. *J. R. Statist. Soc. A* **125** 321–372.

**Basu, D.** (1958). On sampling with and without replacement. *Sankhyā* **20** 287–294.

**Basu, D.** (1964). Recovery of ancillary information. *Sankhyā A* **26** 3–16.

**Basu, D.** (1969). Role of the sufficiency and likelihood principles in sample survey theory. *Sankhyā A* **31** 441–454.

**Basu, D.** (1971). An essay on the logical foundations of survey sampling, part one. In V. P. Godambe and D. A. Sprott, Eds., *Foundations of statistical inference.* Toronto: Holt, Rinehart & Winston, 203–242.

**Basu, D. and J. K. Ghosh** (1967). Sufficient statistics in sampling from a finite universe. *Proc. 36th Session. Int. Statist. Inst.* 850–859.

**Bayless, D. L. and J. N. K. Rao** (1970). An empirical study of stabilities of estimators and variance estimators in unequal probability sampling ($n = 3$ or 4). *J. Amer. Statist. Ass.* **65** 1645–1667.

**Birnbaum, A.** (1962). On the foundations of statistical inference. *J. Amer. Statist. Ass.* **57** 269–326.

**Blackwell, D. and M. A. Girshick** (1954). *Theory of games and statistical decisions.* New York: Wiley.

**Blight, B. J. N. and A. J. Scott** (1973). A stochastic model for repeated surveys. *J. R. Statist. Soc. B* **35** 61–66.

**Brewer, K. R. W.** (1963a). A model of systematic sampling with unequal probabilities. *Aust. J. Statist.* **5** 5–13.

**Brewer, K. R. W.** (1963b). Ratio estimation and finite populations: Some results deducible from the assumption of an underlying stochastic process. *Aust. J. Statist.* **5** 93–105.

**Carroll, J. L. and H. O. Hartley** (1964). The symmetric method of unequal probability sampling without replacement. Abstract, *Biometrics* **20** 908–909.

**Cassel, C. M. and C. E. Särndal** (1972). A model for studying robustness of estimators and informativeness of labels in sampling with varying probabilities. *J. R. Statist. Soc. B* **34** 279–289.

**Cassel, C. M. and C. E. Särndal** (1974). Evaluation of some sampling strategies for finite populations using a continuous variable framework. *Commun. Statist.* **3** 373–390.

**Cassel, C. M., C. E. Särndal, and J. H. Wretman** (1976). Some results on generalized difference estimation and generalized regression estimation for finite populations. *Biometrika* **63** 615–620.

**Chaudhuri, A.** (1969). Minimax solutions of some problems in sampling from a finite population. *Calcutta Statist. Ass. Bull.* **18** 1–24.

**Cochran, W. G.** (1939). The use of analysis of variance in enumeration by sampling. *J. Amer. Statist. Ass.* **34** 492–510.

**Cochran, W. G.** (1946). Relative accuracy of systematic and stratified random samples for a certain class of populations. *Ann. Math. Statist.* **17** 164–177.

**Cochran, W. G.** (1953). *Sampling techniques*, 1st ed. New York: Wiley.

**Cochran, W. G.** (1963). *Sampling techniques*, 2nd ed. New York: Wiley.

**Cox, D. R. and D. V. Hinkley** (1974). *Theoretical statistics.* London: Chapman & Hall.

**Dalenius, T.** (1957). *Sampling in Sweden.* Stockholm: Almqvist & Wiksell.

**David, H. A.** (1970). *Order statistics.* New York: Wiley.

**Deming, W. E. and F. Stephan** (1941). On the interpretation of censuses as samples. *J. Amer. Statist. Ass.* **36** 45–49.

**Durbin, J.** (1967). Estimation of sampling error in multistage surveys. *Appl. Statist.* **16** 152–164.

**Ericson, W. A.** (1965). Optimum stratified sampling using prior information. *J. Amer. Statist. Ass.* **60** 750–771.

**Ericson, W. A.** (1969a). Subjective Bayesian models in sampling finite populations. *J. R. Statist. Soc. B* **31** 195–224.

**Ericson, W. A.** (1969b). Subjective Bayesian models in sampling finite populations: Stratification. In N. L. Johnson and H. Smith, Eds., *New developments in survey sampling*. New York: Wiley–Interscience, 326–357.

**Ericson, W. A.** (1970). On a class of uniformly admissible estimators of a finite population total. *Ann. Math. Statist.* **41** 1369–1372.

**Fellegi, I. P.** (1963). Sampling with varying probabilities without replacement: Rotating and non-rotating samples. *J. Amer. Statist. Ass.* **58** 183–201.

**Finetti, B. de** (1937). La prévision: Ses lois logiques, ses sources subjectives. *Annales de l'Institut Henri Poincaré* **7** 1–68.

**Fisher, R. A.** (1956). *Statistical methods and scientific inference.* London: Oliver & Boyd.

**Foreman, E. K. and K. R. W. Brewer** (1971). The efficient use of supplementary information in standard sampling procedures. *J. R. Statist. Soc. B* **33** 391–400.

**Fraser, D. A. S.** (1957). *Nonparametric methods in statistics.* New York: Wiley.

**Fuller, W. A.** (1970). Simple estimators for the mean of skewed populations. Tech. report, Iowa State University.

**Godambe, V. P.** (1955). A unified theory of sampling from finite populations. *J. R. Statist. Soc. B* **17** 269–278.

**Godambe, V. P.** (1960). An admissible estimate for any sampling design. *Sankhyā* **22** 285–288.

**Godambe, V. P.** (1965). A review of the contributions towards a unified theory of sampling from finite populations. *Rev. Int. Statist. Inst.* **33** 242–258.

**Godambe, V. P.** (1966). A new approach to sampling from finite populations I, II. *J. R. Statist. Soc. B* **28** 310–328.

**Godambe, V. P.** (1968). Bayesian sufficiency in survey sampling. *Ann. Inst. Statist. Math.* **20** 363–373.

**Godambe, V. P.** (1969a). Admissibility and Bayes estimation in sampling finite populations – V. *Ann. Math. Statist.* **40** 672–676.

**Godambe, V. P.** (1969b). A fiducial argument with application to survey sampling. *J. R. Statist. Soc. B* **31** 246–260.

**Godambe, V. P.** (1970). Foundations of survey sampling. *The American Statistician* **24** 33–38.

**Godambe, V. P.** (1975). A reply to my critics. *Sankhyā C* **37** 53–76.

**Godambe, V. P. and V. M. Joshi** (1965). Admissibility and Bayes estimation in sampling finite populations I. *Ann. Math. Statist.* **36** 1707–1722.

**Godambe, V. P. and M. E. Thompson** (1971a). The specification of prior knowledge by classes of prior distributions in survey sampling estimation. In V. P. Godambe and D. A. Sprott, Eds., *Foundations of statistical inference.* Toronto: Holt, Rinehart & Winston, 243–254.

**Godambe, V. P. and M. E. Thompson** (1971b). Bayes, fiducial and frequency aspects of statistical inference in regression analysis in survey sampling. *J. R. Statist. Soc. B* **33** 361–390.

**Godambe, V. P. and M. E. Thompson** (1973). Estimation in sampling theory with exchangeable prior distributions. *Ann. Statist.* **1** 1212–1221.

**Goodman, L. A. and L. Kish** (1950). Controlled selection – a technique in probability sampling. *J. Amer. Statist. Ass.* **45** 350–372.

**Hájek, J.** (1949). Cluster sampling method of two phases. (In Czech.) *Statist. Obzor.* **29** 384–394.

**Hájek, J.** (1959). Optimum strategy and other problems in probability sampling. *Casopis. Pest. Math.* **84** 387–423.

**Hájek, J.** (1964). Asymptotic theory of rejective sampling with varying probabilities from a finite population. *Ann. Math. Statist.* **35** 1491–1523.

**Hansen, M. H. and W. N. Hurwitz** (1943). On the theory of sampling from finite populations. *Ann. Math. Statist.* **14** 333–362.

**Hansen, M. H., W. N. Hurwitz, and L. Pritzker** (1963). The estimation and interpretation of gross differences and the simple response variance. In C. R. Rao, Ed., *Contributions to statistics.* London: Pergamon Press and Calcutta: Statistical Publishing Society.

**Hanurav, T. V.** (1962). On Horvitz and Thompson estimator. *Sankhyā A* **24** 429–436.

**Hanurav, T. V.** (1965). Optimum sampling strategies and some related problems. Ph.D. thesis submitted to the Indian Statistical Institute.

**Hanurav, T. V.** (1966). Some aspects of unified sampling theory. *Sankhyā A* **28** 175–204.

**Hanurav, T. V.** (1967). Optimum utilization of auxiliary information: IIPS sampling of two units from a stratum. *J. R. Statist. Soc. B* **29** 374–391.

**Hanurav, T. V.** (1968). Hyperadmissibility and optimum estimators for sampling finite populations. *Ann. Math. Statist.* **39** 621–642.

**Hartley, H. O. and J. N. K. Rao** (1968). A new estimation theory for sample surveys. *Biometrika* **55** 547–557.

**Hartley, H. O. and J. N. K. Rao** (1969). A new estimation theory for sample surveys, II. In N. L. Johnson and H. Smith, Eds., *New developments in survey sampling.* New York: Wiley–Interscience, 147–169.

**Hartley, H. O. and A. Ross** (1954). Unbiased ratio estimators. *Nature* **174** 270–271.

**Hewitt, E. and L. J. Savage** (1955). Symmetric measures on Cartesian products. *Trans. Amer. Math. Soc.* **80** 470–501.

**Horvitz, D. G. and D. J. Thompson** (1952). A generalisation of sampling without replacement from a finite universe. *J. Amer. Statist. Ass.* **47** 663–685.

**Johnston, J.** (1972). *Econometric methods*, 2nd ed. New York: McGraw-Hill.

**Joshi, V. M.** (1965a). Admissibility and Bayes estimation in sampling finite populations. II. *Ann. Math. Statist.* **36** 1723–1729.

**Joshi, V. M.** (1965b). Admissibility and Bayes estimation in sampling finite populations. III. *Ann. Math. Statist.* **36** 1730–1742.

**Joshi, V. M.** (1966). Admissibility and Bayes estimation in sampling finite populations. IV. *Ann. Math. Statist.* **37** 1658–1670.

**Joshi, V. M.** (1968). Admissibility of the sample mean as estimate of the mean of a finite population. *Ann. Math. Statist.* **39** 606–620.

**Joshi, V. M.** (1969). Admissibility of estimates of the mean of a finite population. In N. L. Johnson and H. Smith, Eds., *New developments in survey sampling.* New York: Wiley–Interscience, 188–212.

**Joshi, V. M.** (1971). Hyperadmissibility of estimators for finite populations. *Ann. Math. Statist.* **42** 680–690.

**Joshi, V. M.** (1972). A note on hyperadmissibility of estimators for finite populations. *Ann. Math. Statist.* **43** 1323–1328.

**Kalbfleisch, J. D. and D. A. Sprott** (1969). Applications of likelihood and fiducial probability to sampling finite populations. In N. L. Johnson and H. Smith, Eds., *New developments in survey sampling.* New York: Wiley–Interscience, 358–389.

**Kempthorne, O.** (1969). Some remarks on statistical inference in finite sampling. In N. L. Johnson and H. Smith, Eds., *New developments in survey sampling.* New York: Wiley–Interscience, 671–695.

**Koop, J. C.** (1963). On the axioms of sample formation and their bearing on the construction of linear estimators in sampling theory for finite universes I, II, III. *Metrika* **7** 81–114, 165–204.

**Koop, J. C.** (1974). Notes for a unified theory of estimation for sample surveys taking into account response errors. *Metrika* **21** 19–39.

**Lahiri, D. B.** (1951). A method of sample selection providing unbiased ratio estimates. *Bull. Int. Statist. Inst.* **33**: II, 133–140.

**Lanke, J.** (1972). Sampling distinguishable elements with replacement. *Ann. Math. Statist.* **43** 1329–1332.

**Lanke, J.** (1973). On UMV-estimators in survey sampling. *Metrika* **20** 196–202.

**Lanke, J.** (1975). *Some contributions to the theory of survey sampling.* Lund: AV-Centralen.

**Lanke, J. and M. K. Ramakrishnan** (1974). Hyperadmissibility in survey sampling. *Ann. Statist.* **2** 205–215.

**Lindgren, B. W.** (1976). *Statistical theory*, 3rd ed. New York: Macmillan.

**Lindley, D. V.** (1971a). Discussion of paper by V. P. Godambe and M. E. Thompson. In V. P. Godambe and D. A. Sprott, Eds., *Foundations of statistical inference.* Toronto: Holt, Rinehart & Winston, 256–257.

**Lindley, D. V.** (1971b). *Bayesian statistics, a review.* Philadelphia: Society for Industrial and Applied Mathematics.

**Lloyd, E. H.** (1952). Least-squares estimation of location and scale parameters using order statistics. *Biometrika* **39** 88–95.

**Madow, W. G.** (1949). On the theory of systematic sampling II. *Ann. Math. Statist.* **20** 333–354.

**Madow, W. G. and M. H. Hansen** (1975). On statistical models and estimation in sample surveys. Contributed papers, 40th Session, Int. Statist. Inst. 554–557.

**Madow, W. G. and L. H. Madow** (1944). On the theory of systematic sampling. *Ann. Math. Statist.* **15** 1–24.

**Mickey, M. R.** (1959). Some finite population unbiased ratio and regression estimators. *J. Amer. Statist. Ass.* **54** 594–612.

**Midzuno, H.** (1952). On the sampling system with probability proportional to sum of sizes. *Ann. Inst. Statist. Math.* **3** 99–107.

**Murthy, M. N.** (1957). Ordered and unordered estimators in sampling without replacement. *Sankhyā* **18** 379–390.

**Murthy M. N.** (1967). *Sampling theory and methods.* Calcutta: Statistical Publishing Society.

**Narain, R. D.** (1951). On sampling without replacement with varying probabilities. *J. Indian Soc. Agric. Statist.* **3** 169–174.

**Neyman, J.** (1934). On the two different aspects of the representative method: The method of stratified sampling and the method of purposive selection. *J. R. Statist. Soc.* **97** 558–625.

**Neyman, J.** (1971). Discussion of paper by R. M. Royall. In V. P. Godambe and D. A. Sprott, Eds., *Foundations of statistical inference.* Toronto: Holt, Rinehart & Winston, 276–278.

**Pathak, P. K.** (1962). On simple random sampling with replacement. *Sankhyā A* **24** 287–302.

**Pascual, J. N.** (1961). Unbiased ratio estimators in stratified sampling. *J. Amer. Statist. Ass.* **56** 70–87.

**Quenouille, M. H.** (1956). Notes on bias in estimation. *Biometrika* **43** 353–360.

**Raiffa, H. and R. O. Schlaifer** (1961). *Applied statistical decision theory.* Boston: Grad. School of Bus. Admin., Harvard University.

**Raj, D.** (1954). Ratio estimation in sampling with equal and unequal probabilities. *J. Indian Soc. Agric. Statist.* **6** 127–138.

**Raj, D.** (1956). Some estimators in sampling with varying probabilities without replacement. *J. Amer. Statist. Ass.* **51** 269–284.

**Raj, D. and S. H. Khamis** (1958). Some remarks on sampling with replacement. *Ann. Math. Statist.* **29** 550–557.

**Ramakrishnan, M. K.** (1970). Optimum estimators and strategies in survey sampling. Ph.D. Thesis, Indian Statistical Institute.

**Ramakrishnan, M. K.** (1975). Choice of an optimum sampling strategy – I. *Ann. Statist.* **3** 669–679.

**Rao, C. R.** (1971). Some aspects of statistical inference in problems of sampling from finite populations. In V. P. Godambe and D. A. Sprott, Eds., *Foundations of statistical inference.* Toronto: Holt, Rinehart & Winston, 177–202.

**Rao, J. N. K.** (1961). On the estimate of the variance in unequal probability sampling. *Ann. Inst. Statist. Math.* **13** 57–60.

**Rao, J. N. K.** (1963a). On two systems of unequal probability sampling without replacement. *Ann. Inst. Statist. Math.* **15** 67–72.

**Rao, J. N. K.** (1963b). On three procedures of unequal probability sampling without replacement. *J. Amer. Statist. Ass.* **58** 202–215.

**Rao, J. N. K.** (1965). On two simple schemes of unequal probability sampling without replacement. *J. Ind. Statist. Ass.* **3** 173–180.

**Rao, J. N. K.** (1966a). Alternative estimators in PPS sampling for multiple characteristics. *Sankhyā A* **28** 47–60.

**Rao, J. N. K.** (1966b). On the relative efficiency of some estimators in PPS sampling for multiple characteristics. *Sankhyā A* **28** 61–70.

**Rao, J. N. K.** (1966c). On the comparison of sampling with and without replacement. *Rev. Int. Statist. Inst.* **34** 125–138.

**Rao, J. N. K.** (1971). Some thoughts on the foundations of survey sampling. *J. Indian Soc. Agric. Statist.* **23** 69–82.

**Rao, J. N. K.** (1975). On the foundations of survey sampling. In J. N. Shrivastava, Ed., *A survey of statistical design and linear models.* The Hague: North Holland, 489–505.

**Rao, J. N. K. and D. L. Bayless** (1969). An empirical study of the stabilities of estimators and variance estimators in unequal probability sampling of two units per stratum. *J. Amer. Statist. Ass.* **64** 540–549.

**Rao, J. N. K. and D. L. Bellhouse** (1976). Optimal estimators of a finite population mean under certain random permutation models, I. Unpublished report.

**Rao, J. N. K., H. O. Hartley, and W. G. Cochran** (1962). On a simple procedure of unequal probability sampling without replacement. *J. R. Statist. Soc. B* **24** 482–491.

**Rao, J. N. K. and M. P. Singh** (1973). On the choice of estimator in survey sampling. *Aust. J. Statist.* **15** 95–104.

**Rao, J. N. K. and J. T. Webster** (1966). On two methods of bias reduction in the estimation of ratios. *Biometrika* **53** 571–577.

**Rao, T. J.** (1972). Horvitz–Thompson and Des Raj estimators revisited. *Aust. J. Statist.* **14** 227–230.

**Rao, T. V. H.** (1962). An existence theorem in sampling theory. *Sankhyā A* **24** 327–330.

**Rao, P. S. R. S.** (1968). On three procedures of sampling from finite populations. *Biometrika* **55** 438–441.

**Rao, P. S. R. S.** (1969). Comparison of four ratio-type estimates under a model. *J. Amer. Statist. Ass.* **64** 574–580.

**Rao, P. S. R. S. and J. N. K. Rao** (1971a). Small sample results for ratio estimators. *Biometrika* **58** 625–630.

**Rao, P. S. R. S. and J. N. K. Rao** (1971b). Small sample results for ratio estimators. Tech. report, Dept. of Statistics, University of Manitoba.

**Roy, J. and J. M. Chakravarti** (1960). Estimating the mean of a finite population. *Ann. Math. Statist.* **31** 392–398.

**Royall, R. M.** (1968). An old approach to finite population sampling theory. *J. Amer. Statist. Ass.* **63** 1269–1279.

**Royall, R. M.** (1970a). On finite population sampling theory under certain linear regression models. *Biometrika* **57** 377–387.

**Royall, R. M.** (1970b). Finite population sampling – on labels in estimation. *Ann. Math. Statist.* **41** 1774–1779.

**Royall, R. M.** (1971a). Linear regression models in finite population sampling theory. In V. P. Godambe and D. A. Sprott, Eds., *Foundations of statistical inference.* Toronto: Holt, Rinehart & Winston, 259–274.

**Royall, R. M.** (1971b). Discussion of paper by D. Basu. In V. P. Godambe and D. A. Sprott, Eds., *Foundations of statistical inference.* Toronto: Holt, Rinehart & Winston, 238–239.

**Royall, R. M.** (1975). The likelihood principle in finite population sampling theory. 40th Session, Int. Statist. Inst.

**Royall, R. M. and J. Herson** (1973a). Robust estimation in finite populations I. *J. Amer. Statist. Ass.* **68** 880–889.

**Royall, R. M. and J. Herson** (1973b). Robust estimation in finite populations II: Stratification on a size variable. *J. Amer. Statist. Ass.* **68** 890–893.

**Sampford, M. R.** (1967). On sampling without replacement with unequal probabilities of selection. *Biometrika* **54** 499–513.

**Sampford, M. R.** (1969). A comparison of some possible methods of sampling from smallish populations, with units of unequal size. In N. L. Johnson and H. Smith, Eds., *New developments in survey sampling*. New York: Wiley–Interscience, 170–187.

**Sampford, M. R.** (1975). The Horvitz–Thompson method in theory and practice – an historical survey. 40th Session, Int. Statist. Inst.

**Sarhan, A. E. and B. G. Greenberg, Eds.,** (1962). *Contributions to order statistics*. New York: Wiley.

**Särndal, C. E.** (1972). Sample survey theory vs. general statistical theory: Estimation of the population mean. *Rev. Int. Statist. Inst.* **40** 1–12.

**Särndal, C. E.** (1976). On uniformly minimum variance estimation in finite populations. *Ann. Statist.* **4** 993–997.

**Scott, A. J.** (1975a). On admissibility and uniform admissibility in finite population sampling. *Ann. Statist.* **2** 489–491.

**Scott, A. J.** (1975b). Some comments on the problem of randomization in surveys. 40th Session, Int. Statist. Inst.

**Scott, A. J. and T. M. F. Smith** (1969a). A note on estimating secondary characters in multivariate surveys. *Sankhyā A* **31** 497–498.

**Scott, A. J. and T. M. F. Smith** (1969b). Estimation in multistage surveys. *J. Amer. Statist. Ass.* **64** 830–840.

**Scott, A. J. and T. M. F. Smith** (1974). Analysis of repeated surveys using time series methods. *J. Amer. Statist. Ass.* **69** 674–678.

**Scott, A. J. and T. M. F. Smith** (1975). Minimax designs for sample surveys. *Biometrika* **62** 353–357.

**Sekkappan, Rm. and M. E. Thompson** (1975). On a class of uniformly admissible estimators for finite populations. *Ann. Statist.* **3** 492–499.

**Sen, A. R.** (1953). On the estimate of the variance in sampling with varying probabilities. *J. Indian Soc. Agric. Statist.* 5 119–127.

**Sen, A. R.** (1955). On the selection of $n$ primary sampling units from a stratum structure ($n \geqslant 2$). *Ann. Math. Statist.* **26** 744–751.

**Seth, G. R.** (1966). On estimators of variance of estimate of population total in varying probabilities. *J. Indian Soc. Agric. Statist.* **18** 52–56.

**Smith, T. M. F.** (1976). Discussion of paper by C. M. Cassel, C. E. Särndal, and J. H. Wretman. *Biometrika* **63** 620.

**Solomon, H. and S. Zacks** (1970). Optimal design of sampling from finite populations: A critical review and indication of new research areas. *J. Amer. Statist. Ass.* **65** 653–677.

**Sukhatme, P. V. and B. V. Sukhatme** (1970). *Sampling theory of surveys with applications.* Bombay: Asia Publishing House.

**Thompson, M. E.** (1971). Discussion of paper by C. R. Rao. In V. P. Godambe and D. A. Sprott, Eds., *Foundations of statistical inference.* Toronto: Holt, Rinehart & Winston, 196–198.

**Thomsen, I.** (1974). Design and estimation problems when estimating a regression coefficient from survey data. Unpublished report.

**Tiao, G. C. and W. Y. Tan** (1965). Bayesian analysis of random effect models in the analysis of variance: I. Posterior distribution of variance components. *Biometrika* **52** 37–53.

**Tin, M.** (1965). Comparison of some ratio estimators. *J. Amer. Statist. Ass.* **60** 294–307.

**Watson, G. S.** (1964). Estimation in finite populations. Unpublished report.

**Wijsman, R. A.** (1973). On the attainment of the Cramér–Rao lower bound. *Ann. Statist.* **1** 538–542.

**Wretman, J. H.** (1970). On inference for finite populations under a superpopulation assumption. (In Swedish.) Research report no. 4, Department of mathematics and statistics. University of Umeå, Sweden.

**Yates, F. and P. M. Grundy** (1953). Selection without replacement from within strata with probability proportional to size. *J. R. Statist. Soc. B* **15** 235–261.

**Zacks, S.** (1969). Bayes sequential designs for sampling finite populations. *J. Amer. Statist. Ass.* **64** 1342–1349.

# Index